T0139306

MULTIMEDIA CONTENT ENCRYPTION

OTHER TELECOMMUNICATIONS BOOKS FROM AUERBACH

**Active and Programmable Networks
for Adaptive Architectures and Services**
Syed Asad Hussain
ISBN: 0-8493-8214-9

**Ad Hoc Mobile Wireless Networks:
Principles, Protocols and Applications**
Subir Kumar Sarkar, T.G. Basavaraju,
and C. Puttamadappa
ISBN: 1-4200-6221-2

**Comprehensive Glossary of Telecom
Abbreviations and Acronyms**
Ali Akbar Arabi
ISBN: 1-4200-5866-5

**Contemporary Coding Techniques and
Applications for Mobile Communications**
Onur Osman and Osman Nuri Ucan
ISBN: 1-4200-5461-9

**Context-Aware Pervasive Systems:
Architectures for a New Breed of
Applications**
Seng Loke
ISBN: 0-8493-7255-0

**Data-driven Block Ciphers for Fast
Telecommunication Systems**
Nikolai Moldovyan and Alexander A. Moldovyan
ISBN: 1-4200-5411-2

**Distributed Antenna Systems:
Open Architecture for Future Wireless
Communications**
Honglin Hu, Yan Zhang, and Jijun Luo
ISBN: 1-4200-4288-2

**Encyclopedia of Wireless and Mobile
Communications**
Borko Furht
ISBN: 1-4200-4326-9

**Handbook of Mobile Broadcasting:
DVB-H, DMB, ISDB-T, AND MEDIAFLO**
Borko Furht and Syed A. Ahson
ISBN: 1-4200-5386-8

The Handbook of Mobile Middleware
Paolo Bellavista and Antonio Corradi
ISBN: 0-8493-3833-6

**The Internet of Things: From RFID
to the Next-Generation Pervasive
Networked Systems**
Lu Yan, Yan Zhang, Laurence T. Yang,
and Huansheng Ning
ISBN: 1-4200-5281-0

**Introduction to Mobile Communication
Technology, Services, Markets**
Tony Wakefield, Dave McNally, David Bowler,
and Alan Mayne
ISBN: 1-4200-4653-5

**Millimeter Wave Technology in Wireless
PAN, LAN, and MAN**
Shao-Qiu Xiao, Ming-Tuo Zhou, and Yan Zhang
ISBN: 0-8493-8227-0

**Mobile WiMAX: Toward Broadband
Wireless Metropolitan Area Networks**
Yan Zhang and Hsiao-Hwa Chen
ISBN: 0-8493-2624-9

**Optical Wireless Communications:
IR for Wireless Connectivity**
Roberto Ramirez-Iniguez, Sevia M. Idrus,
and Ziran Sun
ISBN: 0-8493-7209-7

**Performance Optimization of Digital
Communications Systems**
Vladimir Mitlin
ISBN: 0-8493-6896-0

**Physical Principles of Wireless
Communications**
Victor L. Granatstein
ISBN: 0-8493-3259-1

**Principles of Mobile Computing
and Communications**
Mazliza Othman
ISBN: 1-4200-6158-5

**Resource, Mobility, and Security
Management in Wireless Networks
and Mobile Communications**
Yan Zhang, Honglin Hu, and Masayuki Fujise
ISBN: 0-8493-8036-7

Security in Wireless Mesh Networks
Yan Zhang, Jun Zheng, and Honglin Hu
ISBN: 0-8493-8250-5

**Wireless Ad Hoc Networking:
Personal-Area, Local-Area,
and the Sensory-Area Networks**
Shih-Lin Wu and Yu-Chee Tseng
ISBN: 0-8493-9254-3

**Wireless Mesh Networking:
Architectures, Protocols
and Standards**
Yan Zhang, Jijun Luo, and Honglin Hu
ISBN: 0-8493-7399-9

AUERBACH PUBLICATIONS
www.auerbach-publications.com
To Order Call: 1-800-272-7737 • Fax: 1-800-374-3401
E-mail: orders@crcpress.com

MULTIMEDIA
CONTENT
ENCRYPTION

Techniques and Applications

Shiguo Lian

CRC Press
Taylor & Francis Group
Boca Raton   London   New York

CRC Press is an imprint of the
Taylor & Francis Group, an **informa** business
AN AUERBACH BOOK

Auerbach Publications
Taylor & Francis Group
6000 Broken Sound Parkway NW, Suite 300
Boca Raton, FL 33487-2742

© 2009 by Taylor & Francis Group, LLC
Auerbach is an imprint of Taylor & Francis Group, an Informa business

No claim to original U.S. Government works
Printed in the United States of America on acid-free paper
10 9 8 7 6 5 4 3 2 1

International Standard Book Number-13: 978-1-4200-6527-5 (Hardcover)

Library of Congress Cataloging-in-Publication Data

Lian, Shiguo.
    Multimedia content encryption : techniques and applications / author, Shiguo Lian. -- 1st ed.
        p. cm.
    Includes bibliographical references and index.
    ISBN 978-1-4200-6527-5 (alk. paper)
    1. Multimedia systems--Security measures. 2. Data encryption (Computer science) 3. Data protection. I. Title.

QA76.575.L52 2008
005.8'2--dc22
                                                  2008012301

Visit the Taylor & Francis Web site at
http://www.taylorandfrancis.com

and the Auerbach Web site at
http://www.auerbach-publications.com

Dedication

I dedicate this book to my family: my mother, my father, and my wife.

Contents

Preface

With the wide application of such multimedia data as image, audio or video, media content protection has become more and more necessary and urgent. As one of the means to protect media content, multimedia content encryption has attracted more and more researchers and engineers. Multimedia content encryption focuses on the method that uses the cipher to design a multimedia encryption algorithm or scheme. Thus, it is not to design a cipher, but to design an algorithm or scheme based on the cipher. Until now, few books have mentioned multimedia content encryption, and no books have given a thorough or current description of multimedia content encryption techniques.

The aim of this book is to give a detailed description and show the latest research results on multimedia content encryption. The content includes not only multimedia encryption's brief history, performance requirements and applications but also common and special encryption techniques, attacks, performance evaluations and some other open issues.

From the brief history, you will understand the necessity of multimedia content protection and learn some simple multimedia content encryption algorithms. From the discussion of applications, you will learn about typical applications that are needed to protect multimedia data. In the section on common and special encryption algorithms, you will be informed about various multimedia content encryption algorithms and their performances. From the section on attacks, you will learn some typical attacks that are often used to break multimedia content encryption algorithms, and principles for designing a secure multimedia encryption algorithm. Finally, information about the difficulties in multimedia content encryption and potential research topics is presented in a discussion of open issues.

This book will provide you with useful information whether you are a student, academic researcher, industrial practitioner in a related area, E-commerce professional, IT personnel of institutions or governments, a lawyer or law enforcement personnel.

Acknowledgments

I sincerely thank Professor Zhiquan Wang and Professor Jinsheng Sun, who supervised my work while I was studying for my Ph.D. The Ph.D. study forms the basis of this book. Thanks also to Professor Guanrong Chen, who provided innovative inspiration during the cooperative work.

Precious encouragement was received from my wife Ru Jia. This book is dedicated to her.

The publication of this book could not have been possible without the hard work of those at Taylor & Francis who coordinated the work. My sincere thanks to the project coordinators of Taylor & Francis Group.

The Author

Shiguo Lian earned his Ph.D. degree in multimedia security from Nanjing University of Science and Technology, China, in July 2005. He was a research assistant at City University of Hong Kong from March to June 2004 and has been with France Telecom R&D Beijing since July 2005. He is the author or co-author of more than fifty refereed journal and conference articles covering topics of network security and multimedia content protection, including cryptography, secure p2p content sharing, digital rights management (DRM), image or video encryption, watermarking, digital fingerprinting, and authentication. He has contributed nine chapters to books and holds six filed patents. He is a member of IEEE (the Institute of Electrical and Electronics Engineers), SPIE (The International Society for Optical Engineering), EURASIP (The European Association for Signal Image Processing) and Chinese Image and Graphics Association. He is also a member of IEEE ComSoc Communications & Information Security Technical Committee (CIS TC), IEEE Computer Society's Technical Committee on Security and Privacy and Multimedia Communications Technical Committee (MMTC). He is an editor of the *Journal of Universal Computer Science*, co-editor of a special issue on "Secure Multimedia Communication" in the *Journal of Security and Communication Network*, and co-editor of a special issue on "Multimedia Security in Communication (MUSIC)" in the *Journal of Universal Computer Science* and co-editor of a special issue on "secure multimedia services" in the *Journal of Telecommunication Systems*. He is also a member of the Technical Committees of IEEE ICC2008 CSS Symposium, IEEE GLOBECOM2008 CCNS Symposium, UIC-08, MUSIC'08, CASNET 2008 and SecPri_WiMob 2008. He is also a reviewer for several refereed international journals and conferences.

Chapter 1

Introduction

1.1 Background

With the continuing development of both computer and Internet technology, multimedia data (images, videos, audios, etc.) is being used more and more widely, in applications such as video-on-demand, video conferencing, broadcasting, etc. Now, multimedia data is closely related to many aspects of daily life, including education, commerce, and politics. In order to maintain privacy or security, sensitive data needs to be protected before transmission or distribution. Originally, an access right control method is used, which controls access by authenticating the users. For example, in video-on-demand, a user name and password are used to control the browsing or downloading operations. However, in this method, the multimedia data itself is not protected, and may be stolen during the transmission process. Thus, to maintain security, multimedia data should be protected before transmission or distribution. The typical protection method is the encryption technique [1], which transforms the data from the original form into an unintelligible form.

Until now, various data encryption algorithms have been proposed and widely used, such as AES, RSA, or IDEA [1, 2], most of which are used in text or binary data. It is difficult to use them directly in multimedia data, for multimedia data [3] are often of high redundancy, of large volumes and require real-time interactions, such as displaying, cutting, copying, bit rate conversion, etc. For example, the image shown in Figure 1.1(a) is encrypted into that shown in Figure 1.1(b) by AES algorithm directly. As can be seen, Figure 1.1(b) is still intelligible to some extent. This is because the adjacent pixels in an image are of close relation which cannot be removed by AES algorithm. Besides the security issue, encrypting images or videos with these ciphers directly is time consuming and not suitable for real-time

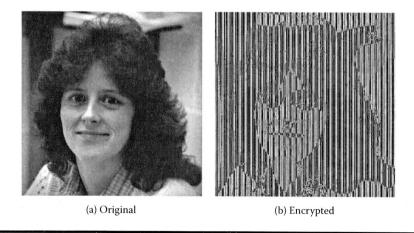

(a) Original   (b) Encrypted

Figure 1.1   The image is encrypted by AES directly.

applications. Therefore, for multimedia data, some new encryption algorithms need to be studied.

1.2 Definition

Multimedia content encryption means to adopt traditional encryption algorithms or novel encryption algorithms to protect multimedia content. Generally, a multimedia encryption system is composed of several components, as shown in Figure 1.2. Here, the original multimedia content is transformed into the encrypted multimedia content with the encryption algorithm under the control of the encryption key. Similarly, the encrypted multimedia content is decrypted into the original multimedia content with the decryption algorithm under the control of the decryption key. Additionally, some attacks may be done to break the system and obtain

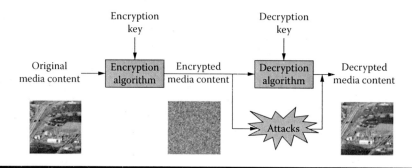

Figure 1.2   Architecture of multimedia encryption and decryption.

the original multimedia content. Most of the research work focuses on efficient encryption and decryption algorithms that are secure against attacks.

1.3 History

Multimedia encryption technology was first reported in the 1980s [4], and became a hot research topic in the second half of the 1990s. Its development can be partitioned into three phases, raw data encryption, compressed data encryption, and partial encryption.

Before the 1990s, few multimedia encoding methods were standardized. Most multimedia data, such as image or video, were stored or transmitted in the raw form. Multimedia encryption was based mostly on pixel scrambling or permutation. That is, the image or video is permuted so that the resulting image is unintelligible. For example, space filling curves [4] are used to permute image or video data, which confuse the relation between adjacent image pixels or video pixels. European TV networks adopt the Eurocrypt [5] standard to encrypt TV signals, which permutes each TV field line by line. These methods are of low computing complexity and low cost. However, the permutation operation changes the relation between adjacent pixels, which will make subsequent compression operations not work. Thus, these encryption algorithms are only suitable for applications that need no compression.

In the first half of the 1990s, with the development of multimedia technology, some image, audio or video encoding standards were developed, such as JPEG, MPEG1/2, etc. Generally, multimedia data are compressed before being stored or transmitted. Thus, the permutation algorithms for raw data encryption are not suitable for these applications. Alternatively, novel algorithms that encrypt the compressed data directly are preferred. For example, Qiao and Nahrstedt [6] proposed the VEA algorithm that uses DES algorithm to encrypt video data. Romeo et al. [7] proposed the RPK algorithm that combines a stream cipher and a block cipher. These algorithms focus on the system security. However, because they encrypt all the compressed data, the computing cost is high, which makes it difficult for large volume data. Additionally, the file format is changed by the encryption algorithm, which means the encrypted data cannot be played or browsed directly. Thus, these algorithms are more suitable for secure multimedia storing than for real-time transmission.

Since the second half of the 1990s, with the development of Internet technology, multimedia applications created more requirements for real-time operation and interaction. By encrypting only parts of the media data, the encrypted data volumes can be greatly reduced, which improves the encryption efficiency. For example, Cheng and Li [8] proposed the algorithm that encrypts only parts of the data stream in wavelet transformed images or videos. Lian et al. [9] proposed an algorithm that encrypts only parts of the parameters in Advance Video Coding, and Servetti et al. [10] proposed an algorithm that encrypts only the bit allocation parameters in MP3 files. These algorithms encrypt the few parts that are significant in human perception, while leaving the other parts unchanged. Thus, the real-time requirement can be met. Additionally, the file format can be kept unchanged, which benefits the transmission process.

1.4 Classification

During the past decades, various multimedia encryption algorithms have been studied, which can be classified into various types.

According to the multimedia content to be encrypted, multimedia encryption algorithms can be classified into three types, image encryption, audio encryption, and video encryption. Generally, for different content, different encryption algorithms should be adopted. In image or video encryption [11–14], the original image or video is transformed into another image or video that is too chaotic to be understood by human eyes. Similarly, in audio encryption [15, 16], the original audio content is transformed into another content that is unintelligible to the human ear.

According to the encryption method, the algorithms can be classified into direct encryption, partial encryption and compression-combined encryption. Generally, different algorithms encrypt different data volumes and thus get different security and efficiency. In direct encryption [6, 17], the multimedia content or compressed content is encrypted with a novel or traditional cipher directly. In partial encryption [18, 19], only some significant parts of the multimedia content are encrypted, while the other parts are left unencrypted. In compression-combined encryption [20, 21], the encryption operation is combined with a compression operation, and they are implemented simultaneously. Intuitively, direct encryption often encrypts the largest data volumes, and thus, is of the highest security and lowest efficiency. Partial encryption and compression-combined encryption reduce the encrypted data volumes, and thus, get higher efficiency and lower security.

According to the properties of the encryption algorithm, they can be classified into perceptual encryption, scalable encryption and so on. Generally, different encryption algorithms have different properties and are suitable for different applications. In perceptual encryption [22, 23], multimedia content is encrypted under the control of the encryption strength that determines the perceptibility of the encrypted multimedia content. A typical case of perceptual encryption is secure multimedia preview, in which the multimedia content is first encrypted with slight encryption strength and decrypted after payment. In scalable encryption [24, 25], the scalable multimedia content is encrypted layer by layer in a progressive manner according to the significance of the layers. It can be used in secure media transcoding. When the encrypted media content is transmitted from the Internet to bandwidth-limited mobile networks, the insignificant layers can be cut off directly without decryption.

1.5 Book Organization

This book covers various aspects of multimedia content encryption. First, the general performance requirements of multimedia content encryption are presented in Chapter 2, and the fundamental techniques of multimedia content encryption are introduced in Chapter 3. Then, some information on encryption techniques follows. The common encryption techniques, including complete encryption, partial encryption and compression-combined encryption, will be investigated, analyzed and evaluated in Chapters 4, 5, and 6. The special encryption techniques, including perceptual encryption, scalable encryption, commutative encryption and watermarking and joint fingerprint embedding and decryption, will be investigated and analyzed in Chapters 7, 8, 9, and 10. Additionally, some attacks on multimedia content encryption are introduced in Chapter 11, and principles for designing secure multimedia content encryption algorithms are proposed in Chapter 12. Furthermore, some typical applications based on multimedia content encryption are presented in Chapter 13. Finally, the open issues and hot topics in multimedia content encryption are proposed in Chapter 14. Chapter 15 is a summary.

References

[1] S. A. Vanstone, A. J. Menezes, and P. C. Oorschot. 1996. *Handbook of Applied Cryptography*. Boca Raton, FL: CRC Press.

[2] R. A. Mollin. 2006. *An Introduction to Cryptography*. Boca Raton, FL: CRC Press.

[3] B. Furht. 1999. *Handbook of Internet and Multimedia Systems and Applications*. Boca Raton, FL: CRC Press.

[4] Y. Matias, and A. Shamir. 1987. A video scrambling technique based on space filling curves. *Proceedings on Advances in Cryptology-CRYPTO'87*, Lecture Notes in Computer Science, Vol. 293, 398–417.

[5] CENELEC (European Committee for Electrotechnical Standardization). 1992 (December). Access control system for the MAC/packet family: EUROCRYPT. European Standard EN 50094. Brussels: CENELEC.

[6] L. Qiao, and K. Nahrstedt. 1997. A new algorithm for MPEG video encryption. In *Proceeding of the First International Conference on Imaging Science, Systems and Technology (CISST'97)*. Las Vegas, NV, July, 21–29.

[7] A. Romeo, G. Romdotti, M. Mattavelli, and D. Mlynek. 1999. Cryptosystem architectures for very high throughput multimedia encryption: The RPK solution. The 6th IEEE International Conference on Electronics, Circuits and Systems, September 5–8. *Proceedings of ICECS '99*, Vol. 1, 261–264.

[8] H. Cheng, and X. Li. 2000. Partial encryption of compressed images and videos. *IEEE Transactions on Signal Processing* 48(8): 2439–2451.

[9] S. Lian, Z. Liu, Z. Ren, and Z. Wang. 2005. Selective video encryption based on advanced video coding. In *Proceedings of 2005 Pacific-Rim Conference on Multimedia (PCM2005)*, Part II, Lecture Notes in Computer Science, Vol. 3768, 281–290.

[10] A. Servetti, C. Testa, J. Carlos, and D. Martin. 2003. Frequency-Selective Partial Encryption of Compressed Audio. Paper presented at the International Conference on Audio, Speech and Signal Processing, Hong Kong, April.

[11] S. Lian, J. Sun, D. Zhang, and Z. Wang. 2004. A Selective Image Encryption Scheme Based on JPEG2000 Codec. Paper presented at 2004 Pacific-Rim Conference on Multimedia (PCM2004), Lecture Notes in Computer Science, Vol. 3332, 65–72.

[12] C. Ho, and W. Hsu. 2005. System and method for image protection. TW227628B.

[13] L. Tang. 1996. Methods for encrypting and decrypting MPEG video data efficiently. In *Proceedings of the Fourth ACM International Multimedia Conference (ACM Multimedia'96)*. Boston, MA, November, 219–230.

[14] C. Shi, and B. Bhargava. 1998. A fast MPEG video encryption algorithm. In *Proceedings of the 6th ACM International Multimedia Conference*. Bristol, UK, September, 81–88.

[15] S. Sridharan, E. Dawson, and B. Goldburg. 1991. Fast Fourier transform based speech encryption system. *IEE Proceedings of Communications, Speech and Vision* 138(3): 215–223.

[16] L. Gang, A. N. Akansu, M. Ramkumar, and X. Xie. 2001. Online music protection and MP3 compression. In *Proceedings of International Symposium on Intelligent Multimedia, Video and Speech Processing*, May, 13–16.

[17] Y. B. Mao, G. R. Chen, and S. G. Lian. 2004. A novel fast image encryption scheme based on the 3D Chaotic Baker Map. *International Journal of Bifurcation and Chaos* 14(10): 3613–3624.

[18] C. Shi, S. Wang, and B. Bhargava. 1999. MPEG video encryption in real-time using secret key cryptography. In *Proceedings of Parallel and Distributed Processing, Technologies and Applications '99*. Las Vegas, NV.

[19] W. Zeng, and S. Lei. 2003. Efficient frequency domain selective scrambling of digital video. *IEEE Transactions on Multimedia* 5(1): 118–129.

[20] A. S. Tosun, and W.-C. Feng. 2001. On error preserving encryption algorithms for wireless video transmission. *Proceedings of the ACM International Multimedia Conference and Exhibition* Vol. 4, Ottawa, Ontario, Canada, 302–308.

[21] C. Wu, and C. C. Jay Kuo. 2001. Efficient multimedia encryption via entropy codec design. Paper presented at SPIE International Symposium on Electronic Imaging 2001, San Jose, CA, Jan. *Proceedings of SPIE* 4314: 128–138.

[22] A. Torrubia, and F. Mora. 2003. Perceptual cryptography of JPEG compressed images on the JFIF bit-stream domain. *Proceedings of the IEEE International Symposium on Consumer Electronics, ISCE*, June 17-19, 58–59.

[23] S. Lian, J. Sun, and Z. Wang. 2004. Perceptual cryptography on SPIHT compressed images or videos. IEEE International Conference on Multimedia and Expo (I) (ICME 2004), June, Vol. 3, 2195–2198.

[24] S. J. Wee, and J. G. Apostolopoulos. 2001. Secure scalable video streaming for wireless networks. In *Proceedings of the IEEE International Conference on Acoustics, Speech, and Signal Processing*. Salt Lake City, UT, May, Vol. 4, 2049–2052.

[25] B. B. Zhu, C. Yuan, Y. Wang, and S. Li. 2005. Scalable protection for MPEG-4 fine granularity scalability. *IEEE Transactions on Multimedia* 7(2): 222–233.

Chapter 2

Performance Requirement of Multimedia Content Encryption

2.1 Introduction

Compared with text or binary data, multimedia data often has high redundancy, large volumes, real-time operations, and compressed data of a certain format. All these properties require that multimedia encryption algorithms satisfy certain requirements. For example, because media content is of large redundancy, directly encrypting it with a traditional cipher may not be secure enough. The relation between encryption and compression should be investigated in order to avoid changes in compression ratio. In some real-time applications, such as Live TV [1] and mobile TV [2], encryption operations should be efficient enough to avoid service delay. Additionally, the encrypted multimedia data may be degraded by transmission errors occurring in interactive services.

Some requirements of multimedia encryption are presented below. They cover various aspects, including security, compression efficiency, encryption efficiency, and format compliance. Additionally, for each performance criterion, some metrics are defined to measure it. These will guide the design of a good multimedia encryption algorithm, and the metrics will provide the means for evaluating various algorithms.

The rest of the chapter is arranged as follows. In Section 2.2, the security requirement is presented together with some metrics. Then, the compression efficiency, encryption efficiency and format compliance are proposed in Sections 2.3, 2.4, and 2.5, respectively. Section 2.6 is a summary.

2.2 Security Requirement

Security is the basic requirement of multimedia content encryption. Different from text/binary encryption, multimedia encryption requires both cryptographic security and perceptual security. The former refers to security against cryptographic attacks [3, 4], and the latter means that the encrypted multimedia content is unintelligible to human perception. In the context of perceptual security, knowing only parts of the multimedia data may be of little help in understanding the multimedia content. Thus, encrypting parts of the multimedia data may be reasonable if the cryptographic security is confirmed. Additionally, for some multimedia encryption applications, the encryption algorithm may be regarded as secure if the cost for breaking it is no smaller than the one paid for the multimedia content. This is different from traditional cipher for text/binary data encryption. For example, in broadcasting, the news may be of no value after an hour. Thus, if the attacker cannot break the encryption algorithm over the course of an hour, then the encryption algorithm may be regarded as secure in this application.

2.2.1 Cryptographic Security

Cryptographic security is determined by the ability to resist the cryptanalysis methods, including such attacks as differential analysis, related-key attack, and statistical attack. Generally, a cipher should be thoroughly analyzed before it can be used in practice. Some simple metrics can be used to measure the cipher's resistance to some typical attacks [4], for example, key sensitivity, plaintext sensitivity, and ciphertext randomness. Generally, if the encryption algorithm is secure against most of the attacks, we say that the encryption algorithm is of high security against cryptographic attacks. Otherwise, the encryption algorithm is regarded as of low security.

2.2.1.1 Key Sensitivity

Key sensitivity is defined as the ciphertext's changes caused by the key's changes. In a good cipher, the slight difference in the keys should cause great changes in the ciphertexts. Set $P = p_0 p_1 \dots p_{n-1}$, $C_0 = c_{0,0} c_{0,1} \dots c_{0,n-1}$, $C_1 = c_{1,0} c_{1,1} \dots c_{1,n-1}$, $K_0 = k_{0,0} k_{0,1} \dots k_{0,n-1}$ and $K_1 = k_{1,0} k_{1,1} \dots k_{1,n-1}$ be the plaintext, ciphertexts and keys, then the key sensitivity (KS) can be computed by

$$KS = \frac{Dif(C_0, C_1)}{n} \times 100\%$$

where *Dif*() is defined as

$$Dif(C_0, C_1) = \sum_{i=0}^{n-1} (c_{0,i} \oplus c_{1,i}).$$

Here, $\oplus$ is the bit-XOR operation and C_0 and C_1 are computed by

$$\begin{cases} C_0 = E(P, K_0) \\ C_1 = E(P, K_1) \\ Dif(K_0, K_1) = 1/n \end{cases}.$$

Here, there is only one bit of difference between K_0 and K_1. Generally, for a good cipher, the value of KS is about 50%.

2.2.1.2 Plaintext Sensitivity

Similar to key sensitivity, plaintext sensitivity is defined as the changes in the ciphertext changes caused by the changes in the plaintext. In a good cipher, the slight difference in the plaintexts should cause great changes in the ciphertexts. Set $P_0 = p_{0,0} p_{0,1} \cdots p_{0,n-1}$, $P_1 = p_{1,0} p_{1,1} \cdots p_{1,n-1}$, $C_0 = c_{0,0} c_{0,1} \cdots c_{0,n-1}$, $C_1 = c_{1,0} c_{1,1} \cdots c_{1,n-1}$ and $K = k_0 k_1 \cdots k_{n-1}$ be the plaintexts, ciphertexts and key, then the plaintext sensitivity (PS) can be computed by

$$PS = \frac{Dif(C_0, C_1)}{n} \times 100\%,$$

where *Dif*() is defined as

$$Dif(C_0, C_1) = \sum_{i=0}^{n-1} (c_{0,i} \oplus c_{1,i}).$$

Here, $\oplus$ is the bit-XOR operation and C_0 and C_1 are computed by

$$\begin{cases} C_0 = E(P_0, K) \\ C_1 = E(P_1, K) \\ Dif(P_0, P_1) = 1/n \end{cases}.$$

Here, there is only one bit of difference between P_0 and P_1. Generally, for a good cipher, the value of PS is about 50%.

2.2.1.3 Ciphertext Randomness

Generally, the ciphertext is quite different from the plaintext. As a good cipher, the ciphertext often has such good randomness that makes it difficult for attackers to find holes in the ciphertext's statistical properties. Some metrics have been reported for measuring randomness of a sequence [5]. Taking image encryption for example, the histogram [6] can be used to tell the pixel distribution of a cipher-image.

Set $P = p_0 p_1 \ldots p_{n-1}$ and $C = c_0 c_1 \ldots c_{n-1}$ ($0 \le p_i, c_i \le L - 1, i = 0, 1, \ldots, n - 1$) be the plaintext and ciphertext, then the ciphertext's histogram sequence $H = h_0 h_1 \ldots h_{L-1}$ is computed by the following method.

```
For j = 0 to L - 1
   h_j = 0;
End
For i = 0 to n - 1
   h_{c_i} = h_{c_i} + 1;
End
```

The plaintext's histogram can be computed by the same method. Generally, the ciphertext's histogram sequence is more similar to uniform distribution compared with the plaintext's histogram sequence. An example is shown in Figure 2.1, where the image is encrypted by AES. As can be seen, the histogram of the cipher-image is closer to uniform distribution than the plain-image histogram.

2.2.2 Perceptual Security

As has been pointed out, multimedia content encryption is different from text/binary data encryption. There is much more redundancy in multimedia content, which may make the encrypted content still understandable to some extent. Because perceptual security refers to the ciphertext's intelligibility, it can be measured by both subjective and objective metrics.

2.2.2.1 Subjective Metric

In a subjective metric, given ciphertexts with different quality, scorers are invited to comment on the ciphertexts' quality level. The typical quality levels for a subjective metric are listed in Table 2.1. Here, QL0 denotes that the ciphertext's content is completely understandable, even is of good quality. In this case, the encryption is not successful. QL1 denotes that only a few parts of the ciphertext's content can be

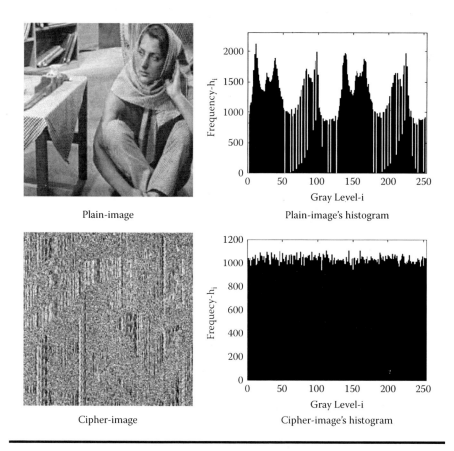

Figure 2.1   Histograms of plain image and cipher image.

understood. For example, only the shape is intelligible, while the texture is unclear. In this case, the encryption is only suitable for some applications not requiring high security. QL2 denotes that the ciphertext's content is completely unintelligible. In this case, the encryption is successful in perceptual aspects. Figure 2.2 shows encrypted images corresponding to different levels.

Table 2.1   Quality Level in a Subjective Metric

Quality Level	Corresponding Quality of Ciphertext
QL0	Completely understandable
QL1	Understandable
QL2	Not understandable

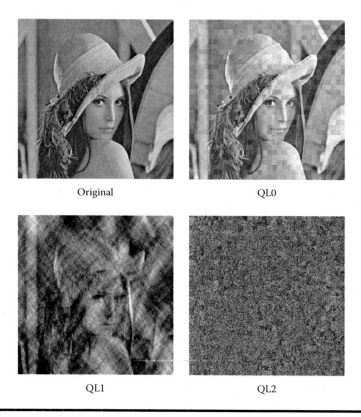

Original	QL0
QL1	QL2

Figure 2.2   Ciphertexts corresponding to different quality levels.

2.2.2.2 Objective Metric

An objective metric provides more efficient test methods and is suitable for computer-based analysis. Intuitively, it should depend on multimedia understanding techniques, for example, image understanding. However, until now, no suitable metric for multimedia content has been reported, except some metrics for multimedia quality. Thus, the quality metrics can be used as the objective metric of ciphertext. The typical metric for audio quality is signal-to-noise ratio (SNR) [7], and the one for image is peak signal-to-noise ratio (PSNR) [8]. Taking PSNR, for example, the computing method is introduced as follows.

PSNR is initially used to measure images' quality losses caused by such operations as compression, noising, transmission errors, etc. It is computed by comparing the original image and the operated image. Here, set $P = p_0 p_1 \ldots p_{n-1}$ and $C = c_0 c_1 \ldots c_{n-1}$ $(0 \leq p_i, c_i \leq L - 1, i = 0, 1, \ldots, n - 1)$ be the plain-image and cipher-image, then the cipher-image's PSNR can be computed by

$$PSNR = 10 \log_{10} \frac{L^2}{MSE},$$

where MSE (mean square error) satisfies

$$MSE = \frac{1}{n} \sum_{i=0}^{n-1} (c_i - p_i)^2.$$

Here, L is the image pixel's gray level. For example, for an 8-bit image, L = 256. Generally, the bigger PSNR is, the higher the cipher-image quality is. Thus, for a good image encryption algorithm, the cipher-image's PSNR should be small enough for context protection.

Fractal dimension (FD) [9–11] is another metric that can be used to measure the quality of the encrypted multimedia content. It is originally used as the extracted feature in multimedia understanding. Taking an image for example, the width and height are regarded as two dimensions, and the pixel's gray level can be regarded as the third dimension. Because the pixels often have different gray levels, the computed FD often ranges from 2 to 3. According to FD computing [9], the more random the image pixels are, the bigger (closer to 3) the computed FD is. Considering that there are similarities between adjacent image pixels, the FD of the original image is often far from 3. Differently, after encryption, the image pixels become random, and thus, the FD of the cipher-image is closer to 3. Figure 2.3 shows the original image and cipher-image. As can be seen, the cipher-image with a bigger FD looks more random than the original one. Thus, as a good multimedia encryption algorithm, it should obtain the ciphertext with a big FD.

2.2.3 Security Level

According to cryptographic security and perceptual security, the algorithms can be classified into three levels, as shown in Table 2.2. Here, SL0 denotes that the encryption algorithm obtains high security in both cryptographic aspect and perceptual aspect (QL2). Encryption algorithms with this level of security are suitable for applications requiring high security. SL1 denotes that the encryption algorithm has high security in the cryptographic aspect and low security in the perceptual aspect. This kind of algorithm should be improved. SL2 denotes that the encryption algorithm has low security in the cryptographic aspect and high security in the perceptual aspect (QL2). Encryption algorithms with this level of security are suitable for applications requiring low security along with good performance in other aspects, such as high efficiency and good error-robustness. SL3 denotes that the encryption algorithm has high cryptographic security and low perceptual security (QL0 or QL1), or has low cryptographic security and low perceptual security (QL0 or QL1). Encryption algorithms with this level of security should be improved before practical applications. Additionally, the power cost is closely related to the security of the encryption algorithm [12–15]. Generally, the greater the cipher's computational complexity, the higher the power cost, and the higher the cipher's security.

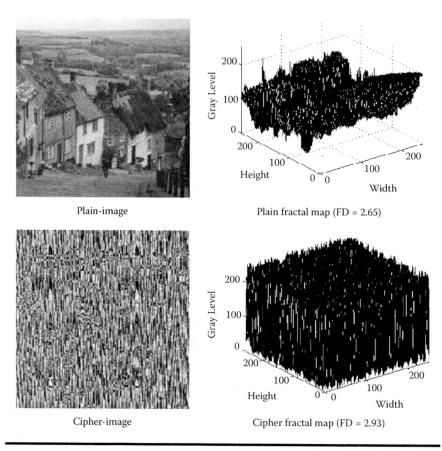

Figure 2.3   Fractal dimensions of the original image and cipher image.

Table 2.2   Security Level of Different Encryption Algorithms

Security Level	Corresponding Performances
SL0	High cryptographic security + high perceptual security (QL2)
SL1	High cryptographic security + low perceptual security (QL0 or QL1)
SL2	Low cryptographic security + high perceptual security (QL2)
SL3	Low cryptographic security + low perceptual security (QL0 or QL1)

2.3 Compression Efficiency

Multimedia data are compressed to reduce the storage space or transmission bandwidth. Multimedia encryption may be applied to multimedia data before compression, during compression or after compression, depending on the applications. However, in all cases, multimedia encryption algorithms should not change the compression ratio or should at least keep the changes in a small range. This is especially important in wireless or mobile applications, in which the channel bandwidth is limited. To evaluate the effects on compression ratio, we define the changed compression ratio (CCR) as the ratio between the changed data size and the original data size. Set the data size as R_0 when multimedia data are not encrypted, and R_1 when multimedia data are encrypted. Then, CCR is defined as

$$CCR = \frac{|R_1 - R_0|}{R_0} \times 100\%.$$

According to the computed *CCR*, we can classify the encryption algorithm into three types, as shown in Table 2.3. CL0 denotes that the compression ratio is unchanged ($CCR = 0$). CL1 denotes that the compression ratio is changed no more than 10% ($CCR \leq 10\%$). CL2 denotes that the compression ratio is changed greatly ($CCR > 10\%$).

2.4 Encryption Efficiency

Because real-time transmission or access is often required by multimedia applications, multimedia encryption algorithms should be efficient so that they don't delay the transmission or access operations. Generally, two kinds of method can be adopted: the first is to reduce the encrypted data volume, and the other is to adopt lightweight encryption algorithms. Considering that multimedia encryption is often used together with multimedia compression, we use the comparative ratio between encryption and compression to measure the encryption speed.

Table 2.3 Compression Level Classification According to CCR

Compression Level	Corresponding CCR
CL0	$CCR = 0$
CL1	$CCR \leq 10\%$
CL2	$CCR > 10\%$

Table 2.4   Encryption Efficiency Level Classification According to CCR

Efficiency Level	Corresponding CCR
EL0	ETR ≤ 10%
EL1	10% < ETR ≤ 50%
EL2	ETR > 50%

Let T_{CE}, T_{CD}, T_E and T_D be the time of compression, decompression, encryption and decryption, respectively. The encryption time ratio (ETR) is defined as

$$ETR = \frac{T_E + T_D}{T_{CE} + T_{CD}} \times 100\%.$$

According to the computed ETR, we can classify the encryption algorithm into three levels, as shown in Table 2.4. EL0 denotes that the encryption/decryption operation is very efficient compared with the compression/decompression operation ($ETR \leq 10\%$), and does not affect real-time applications. EL1 denotes that the encryption/decryption operation is efficient compared with the compression/decompression operation ($10\% < ETR \leq 50\%$), but it is not suitable for real-time applications. EL2 denotes that the encryption/decryption operation is time consuming ($ETR > 50\%$).

2.5 Format Compliance

Multimedia data are often encoded or compressed before transmission, which produces data streams with some format information, such as file header, time stamp, file tail, etc. The format information will be used by the decoders to recover the multimedia data successfully. For example, the format information can be used as the synchronization information in multimedia communication. The encryption operation may be done before, during or after compression, which may have different effects on the multimedia format. According to whether they affect the multimedia format, the encryption algorithms can be classified into three levels, as shown in Table 2.5.

FL0 denotes that the multimedia format is completely changed by the encryption operation. Thus, the encrypted multimedia content cannot be operated by such simple operations as decoding, displaying, editing, etc. In this case, the encryption operation is incompatable with the compression operation.

FL1 denotes that the format information is left unencrypted, and the encrypted multimedia content can be operated by such simple operations as decoding, displaying,

Table 2.5   Encryption Classification According to Format Compliance

Format Level	Corresponding CCR
FL0	Incompliant
FL1	Keep synchronization
FL2	Support direct operations

cutting, pasting, etc. Encrypting the content except the format information makes the encrypted data stream format-compliant, and it also keeps synchronization for communication and improves the data stream's error robustness to some extent.

FL2 denotes that the encrypted multimedia content can be operated directly without decryption. In some applications, it will save on cost to operate directly on the encrypted multimedia data, but not to do the triple operations of decryption-operation-encryption. For example, the encrypted multimedia data can be recompressed, the bit rate of the encrypted multimedia data can be controlled, the image block or frame can be cut, copied or inserted, etc. If the encrypted multimedia data can be operated by certain operations, then the corresponding encryption algorithm supports the corresponding operations.

2.6 Application Suitability

As is known, there exist various multimedia-related applications, such as multimedia storage, multimedia transmission, real-time interaction and wireless/mobile communication. Generally, different applications have different performance requirements. Additionally, there are some contradictions between different performances.

In multimedia storage, multimedia data can first be encrypted then stored, or first stored then decrypted. Since the encryption or decryption operation can be implemented offline, the encryption algorithm's efficiency is not so important. Additionally, media data may be stored on hard disks, such as VCD, DVD, or CD, and slight changes in data size may be acceptable. However, considering that media data are stored for a long time, which provides enough time for attackers, multimedia encryption algorithms should be secure enough. Therefore, in the scenario of multimedia storage, the encryption algorithm with high security is preferred.

In multimedia transmission, media data are delivered from the sender to the receiver after they are encrypted, or they are decrypted after being received. Since the encryption/decryption operation may be done online or offline, the encryption efficiency may be high or low. However, the transmission bandwidth is often limited, which requires that multimedia encryption does not change compression ratio apparently. Thus, in the scenario of multimedia transmission, the encryption algorithm with low CCR is preferred.

In real-time interaction, multimedia data are often encrypted, transmitted and decrypted online in a real-time manner. Compared with general multimedia transmission, real-time interaction requires that the encryption or decryption algorithms should have high encryption efficiency in order to avoid service delay. Thus, in the scenario of real-time interaction, the encryption/decryption algorithm with high encryption efficiency and low CCR is preferred.

In wireless/mobile communication, transmission errors often happen in wireless channels, the transmission bandwidth is now still limited, and the mobile terminal's computational capability is often limited by the power. Thus, compared with general multimedia transmission, wireless/mobile multimedia requires that the encryption algorithm should keep the synchronization information in order to resist transmission errors, not change data size apparently, and not cost much power. Generally, the higher the cipher's security is, the more the cipher's computational complexity is, and the more the power costs. Thus, in the scenario of wireless/mobile communication, the encryption/decryption operation with format compliance, low CCR, high encryption efficiency and low power cost is preferred.

2.7 Summary

In this chapter, performance requirements for multimedia encryption algorithms are presented, together with some metrics. Different performances are required by different applications, and the general performances include cryptographic security, perceptual security, compression efficiency, encryption efficiency, and format compliance. To test these performances, some metrics are defined, including key/plaintext sensitivity, PSNR, changed compression ratio, encryption time ratio, etc. They will be used to guide the design of multimedia encryption algorithms and also to evaluate the encryption algorithms.

References

[1] Live TV. http://en.wikipedia.org/wiki/Live_TV
[2] European Telecommunications Standards Institute (ETSI). 2004 (November). *Digital Video Broadcasting (DVB), Transmission System for Handheld Terminals (DVB-H)*. ETSI, November 2004.
[3] R. A. Mollin. 2006. *An Introduction to Cryptography*. Boca Raton, FL: CRC Press.
[4] William F. Friedman, 1993. *Military Cryptanalysis: Transposition and Fractionating Systems*. Walnut Creek, CA: Aegean Park Press.
[5] National Institute of Standards and Technology (NIST). 2001 (May 25). *Security Requirements for Cryptographic Modules (Change Notice)*. Federal Information Processing Standards Publication (FIPS PUB) 140-1. Gaithersburg, MD: NIST.

[6] A. Marion. 1991. *An Introduction to Image Processing.* London: Chapman and Hall.

[7] Signal-to-Noise Ratio (SNR). http://en.wikipedia.org/wiki/Signal-to-noise_ratio

[8] Peak Signal-to-Noise Ratio (PSNR). http://en.wikipedia.org/wiki/Peak_signal-to-noise_ratio

[9] C. C. Chen, J. S. Daponte, and M. D. Fox. 1989. Fractal feature analysis and classification in medical imaging. *IEEE Transactions on Medical Imaging* 8(2): 133–142.

[10] P. Soille, and J.-F. Rivest. 1996. On the validity of fractal dimension measurements in image analysis. *Journal of Visual Communication and Image Representation* 7: 217–229.

[11] B. Dubuc, J. F. Quiniou, C. Roques-Carmes, C. Tricot, and S. W. Zucker. 1989. Evaluating the fractal dimension of profiles. *Physics Review A*, 39: 1500–1512.

[12] N. R. Potlapally, S. R. A. Raghunathan, and N. K. Jha. 2003. Analyzing the energy consumption of security protocols. In *Proceedings of the 2003 International Symposium on Low Power Electronics and Design*, Seoul, Korea, 30–35.

[13] J. Goodman, and A. P. Chandrakasan. 1998. Low power scalable encryption for wireless systems. *Wireless Networks* 4: 55–70.

[14] K. Tikkanen, M. Hannikainen, T. Hamalainen, and J. Saarinen. 2002. Hardware implementation of the improved WEP and RC4 encryption algorithms for wireless terminals. In *Proceedings European Signal Processing Conference*, 2289–2292.

[15] Chui Sian Ong, Klara Nahrstedt, and Wanghong Yuan. 2003. Quality of protection for mobile multimedia applications. In *Proceedings IEEE International Conference on Multimedia and Expo (ICME2003)*, Baltimore, MD, July.

Chapter 3

Fundamental Techniques

3.1 Introduction

Multimedia encryption techniques are closely related to some other techniques, such as encryption techniques [1], multimedia compression [2], multimedia communication [3], and digital watermarking [4], etc. First, multimedia encryption aims to encrypt multimedia content with encryption techniques, and thus, multimedia encryption is based on traditional encryption techniques. Second, multimedia content is often compressed before transmission or storage in order to save cost in space or bandwidth, and thus, multimedia encryption should consider the compression operations, for example, before compression, during compression or after compression. Third, multimedia content is often transmitted from the sender to the receiver through multimedia communication techniques, and thus, the multimedia encryption should satisfy different applications in multimedia communication. Fourth, multimedia encryption is often combined with other techniques, such as digital watermarking or fingerprinting, in order to protect some other properties of multimedia content, for example, integrity, ownership, copyright, etc.

These techniques, including encryption techniques, multimedia compression, multimedia communication, and digital watermarking, are introduced in this chapter. Some basic concepts are explained in brief, some terms are introduced with references, and some architectures are described in detail.

3.2 Cryptography

Cryptography [1] provides various means to confirm content security in communication. The means include cipher, hash, digital signature, key generation, and authentication, etc. Among them, the cipher transforms the original data into an

21

unintelligible form under the control of the key, and it is often used to protect the confidentiality of data. The hash generates a short string from the original data, which is often used to protect the integrity of the data. Digital signature uses key-based hash to generate a hash value for the data, and it is often used to detect whether the operation is done by the correct owner or not. Key generation and authentication provides some methods to generate and distribute multiple keys in a communication environment.

Cryptanalysis focuses on the methods to analyze or break cryptographic means. It provides some common or special methods to analyze the security of a cipher, hash, digital signature or key generation/authentication algorithm. Generally, a cryptographic method should survive all the cryptanalysis methods before it can be used in practice.

The cipher and cryptanalysis are briefly introduced below. Additionally, the encryption mode that defines how to use the cipher to encrypt the data will also be reviewed.

3.2.1 Cipher

A cipher transforms the plaintext into ciphertext and recovers the plaintext from ciphertext under control of the key. Here, the transforming operation and the recovering operation are named encryption and decryption, respectively. The key is necessary to decryption, because without the key, the plaintext cannot be recovered correctly. Various ciphers have been reported to date, and some of them have been standardized. According to their properties, the ciphers can be classified into different types.

3.2.1.1 Symmetric Cipher or Asymmetric Cipher

Generally, the decryption key can be same as the encryption key or different from the encryption key. In the former case, the decryption operation is symmetric to the encryption operation, as shown in Figure 3.1(a). They are defined as

$$\begin{cases} C = \mathrm{E}(P,K) \\ P = \mathrm{D}(C,K) \end{cases}.$$

Here, P, $E()$, K, C and $D()$ are the plaintext, encryption operation, key, ciphertext and decryption operation, respectively. Thus, this kind of cipher is termed a symmetric cipher. In the second case, the decryption operation is not symmetric to the encryption operation, as shown in Figure 3.1(b). They are defined as

$$\begin{cases} C = \mathrm{E}(P,K_E) \\ P = \mathrm{D}(C,K_D) \end{cases}.$$

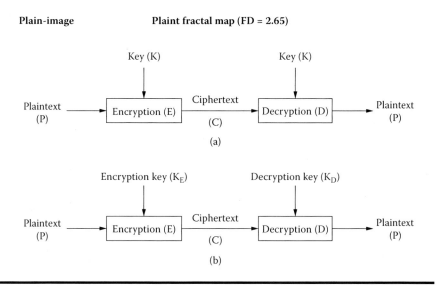

Figure 3.1   Architectures of symmetric cipher and asymmetric cipher.

Here, K_E and K_D are the encryption key and decryption key, respectively. This kind of cipher is termed an asymmetric cipher.

For a symmetric cipher, the encryption key is the same as the decryption key. Thus, the key K is only known to the sender and receiver, and it should be kept private and not be made known to a third party. Otherwise, the ciphertext can be decrypted and released by the third party. According to this property, a symmetric cipher is also called a private cipher. Well-known symmetric ciphers include Data Encryption Standard (DES) [5], Advanced Encryption Standard (AES) [6], International Data Encryption Algorithm (IDEA) [7], etc. These are explained in the next section.

For an asymmetric cipher, the decryption key is different from the encryption key. Thus, the encryption key K_E can be made public, while the decryption key K_D is kept private (only known to the receiver). If the sender or a third party knows only the encryption key, he cannot decrypt the ciphertext. According to this property, an asymmetric cipher is also called a public cipher. Compared with a symmetric or private cipher, an asymmetric or public cipher is more suitable for key exchange in a communication environment. Some public ciphers have been reported, most of which are based on mathematical difficulties. For example, RSA [8] cipher is based on the difficulty of factorization of large prime numbers, Elliptic Curve Cryptography (ECC) [9] is based on the difficulty of the discrete logarithm problem, and ElGamal encryption [10] is defined over any cyclic group; its security depends on the difficulty of a certain problem related to computing discrete logarithms.

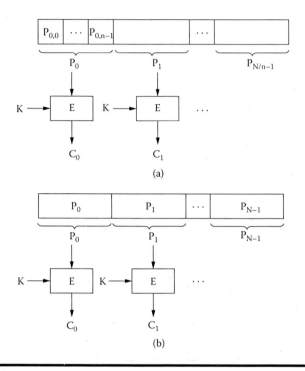

Figure 3.2 Architectures of block cipher and stream cipher.

3.2.1.2 Block Cipher or Stream Cipher

Generally, the original data composed of N ($N > 0$) bits is partitioned into segments, and all the segments are encrypted in order. According to the size of the segment, the cipher can be classified into two types, block cipher and stream cipher. In the former, the segment is a block composed of n ($n > 0$) bits, as shown in Figure 3.2(a). Each block P_i ($i = 0, 1, \ldots, N/n - 1$) is encrypted and decrypted according to

$$\begin{cases} C_i = E(P_i, K) \\ P_i = D(C_i, K) \end{cases} (i = 0, 1, \cdots, N/n - 1).$$

Here, each block P_i is composed of n bits, that is, $p_{i,0} p_{i,1} \cdots p_{i,n-1}$. In a stream cipher, the plaintext is encrypted bit by bit, as shown in Figure 3.2(b). The encryption and decryption operations are defined as

$$\begin{cases} c_i = E(p_i, k_i) \\ p_i = D(c_i, k_i) \end{cases} (i = 0, 1, \cdots, N).$$

Table 3.1   Some Typical Block Ciphers

Block Cipher	Key Size (bits)	Block Size (bits)	Year
DES	56	64	1976
3DES	112 or 168	64	1978
AES	128, 192 or 256	128	1998
RC6	Minimum 0; maximum 2040, multiple of 8 bits; default 128 bits	128	1998
Blowfish	Minimum 32, maximum 448, multiple of 8 bits; default 128 bits	64	1994
CAST-256	Minimum 128, maximum 256, multiple of 32 bits; default 128 bits	128	1998

Here, the bit operation, such as XOR, is often used as the encryption/decryption operation.

Many block ciphers have been reported. Most of them are based on computing security. That is, the encryption or decryption operation is realized by complex computing. The high computing complexity strengthens the cipher system and improves the difficulty for an attacker. Some typical block ciphers [1] are listed in Table 3.1. As can be seen, the block size of plaintext or ciphertext is 64 or 128 (bits). The key size is often a multiple of 8 (bits). Among them, 3DES [11] and AES are used more widely than others.

In stream cipher, the plaintext P is modulated by the random key sequence K. The XOR operation often acts as the modulation operation. The research topic is how to generate the random key sequence. Some random sequence generators have been reported, such as linear feedback shift register (LFSR) [12], the generator based on block cipher [13], or the chaos-based generator [14]. In these generators, the initial vector is required, which together with the key controls the random sequence generation process. Some typical stream ciphers are listed in Table 3.2. Such ciphers as RC4 [15], SNOW [16] or Rabbit [17] are often used in practice.

Table 3.2   Some Typical Stream Ciphers

Stream Cipher	Key Size (bits)	Initial Vector Size (bits)	Year
RC4	8-2048	8	1987
A5/1	54	114	1989
SNOW	128 or 256	32	2003
Grain	80	64	2004
HC-256	256	256	2004
Rabbit	128	64	2004

3.2.2 Encryption Mode of Block Cipher

Because block ciphers operate on the basis of blocks of a certain length (e.g., 64 bits or 128 bits), the encryption modes are defined to encrypt plaintext of variable length. According to the connection between adjacent plaintexts, ciphertexts and keys, the encryption mode is classified into several types. The typical modes [18] include Electronic Codebook (ECB), Cipher-Block Chaining (CBC), Cipher Feedback (CFB), Output Feedback (OFB), Counter (CTR), etc.

3.2.2.1 ECB Mode

ECB is the simplest encryption mode, as shown in Figure 3.3(a). The plaintext is partitioned into n blocks, and each block is encrypted independently. The decryption process is symmetric to the encryption process, and they are defined as

$$\begin{cases} C_i = \mathrm{E}(P_i, K) \\ P_i = \mathrm{D}(C_i, K) \end{cases} (i = 0,1,\cdots,n-1).$$

In this mode, the identical plain-blocks are encrypted into the same cipher-blocks.

3.2.2.2 CBC Mode

In the CBC mode, as shown in Figure 3.3(b), the ith plain-block is modulated by the $i - 1$th cipher-block before it is encrypted by the block cipher. The first plain-block is modulated by the initial vector (IV). The decryption process is symmetric to the encryption process, and they are defined as

$$\begin{cases} C_i = \mathrm{E}(P_i \oplus C_{i-1}, K), C_{-1} = IV \\ P_i = \mathrm{D}(C_i, K) \oplus C_{i-1}, C_{-1} = IV \end{cases} (i = 0,1,\cdots,n-1).$$

Thus, the identical plain-blocks may be encrypted into different cipher-blocks.

3.2.2.3 CFB Mode

In CFB mode, as shown in Figure 3.3(c), the ith plain-block is modulated by the $i - 1$th cipher-block's encryption result. The first plain-block is modulated by the IV's encryption result. The decryption process is symmetric to the encryption process, and they are defined as

$$\begin{cases} C_i = \mathrm{E}(C_{i-1}, K) \oplus P_i, C_{-1} = IV \\ P_i = \mathrm{E}(C_{i-1}, K) \oplus C_i, C_{-1} = IV \end{cases} (i = 0,1,\cdots,n-1).$$

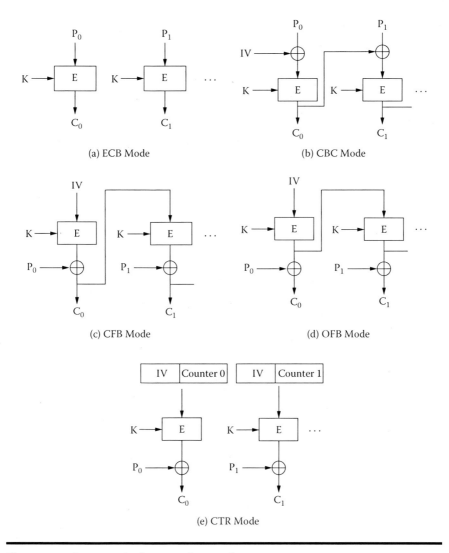

Figure 3.3   Some typical encryption modes.

Thus, the encryption mode is similar to a stream cipher, and the identical plain-blocks may be encrypted into different cipher-blocks.

3.2.2.4 OFB Mode

The OFB mode, as shown in Figure 3.3(d), is similar to the CFB mode, which encrypts plain-blocks in a mode similar to a stream cipher. The difference is that the block cipher's output is feedback instead of the cipher-block. The encryption and decryption operations are defined as

$$\begin{cases} C_i = E^i(IV, K) \oplus P_i \\ P_i = E^i(IV, K) \oplus C_i \end{cases} \quad (i = 0, 1, \cdots, n-1).$$

Here, $E^i(IV,K)$ means to encrypt IV with K for i times. Thus, the identical plain-blocks may be encrypted into different cipher-blocks.

3.2.2.5 CTR Mode

In CTR mode, as shown in Figure 3.3(e), the plain-blocks are encrypted with the mode similar to a stream cipher. Different from the CFB and OFB modes, the CTR mode modulates each plain-block with the result produced by encrypting the IV and a Counter. Here, the Counter changes with i. The encrypton and decryption operations are defined as

$$\begin{cases} C_i = E(IV \| \text{Counter } i, K) \oplus P_i \\ P_i = E(IV \| \text{Counter } i, K) \oplus C_i \end{cases} \quad (i = 0, 1, \cdots, n-1).$$

Since the Counter changes with i, the identical plain-blocks may be encrypted into different cipher-blocks.

3.2.3 Cryptanalysis

Cryptanalysis [19] provides the methods for breaking the cipher system. Generally, for a cipher system, the Kerckhoffs principle is satisfied: the attacker knows the cipher itself, and the cipher's security is determined by the private key. Thus, the analysis methods aim to obtain the cipher's private key under the condition of knowing such information as ciphertext, plaintext, encryption algorithm, etc. According to the information known by attackers, the cryptanalysis method can be classified into various types.

- Ciphertext-only attack denotes the attack method works when the attacker knows only a collection of ciphertexts.
- Known-plaintext attack denotes the attack method works when the attacker has a set of plaintext-ciphertext pairs.
- Chosen-plaintext attack denotes the attack method works when the attacker can obtain the ciphertexts corresponding to an arbitrary set of plaintexts.
- Related-key attack denotes the attack method works when the attacker can obtain the ciphertexts encrypted under two different keys.

The difficulty of the attacks listed decreases in order. For example, the cipher is easier to be broken by related-key attack than by ciphertext-only attack.

According to the attack method, the attacks can be classified into differential analysis, linear analysis, integral analysis, related-key attack, etc.

- Differential analysis [19] is a widely used cryptanalysis method used to break block ciphers, stream ciphers and hash functions. Generally, it studies the relation between the differences in an input and the differences in the corresponding output. In block cipher analysis, it exploits the differences caused by iterated transformations, discovers some apparent behaviors, and recovers the private key.
- Linear analysis is another widely used cryptanalysis method used to block ciphers and stream ciphers. It tries to find affine approximations to the action of a cipher.
- Integral analysis is an attack that is particularly used to break block ciphers based on substitution-permutation networks. It is also named square attack because it is originally a dedicated attack against square.
- Related-key attack [20] denotes any form of analysis method under the condition that the attacker knows the relation between different keys. It aims to obtain the private keys by analyzing the plaintext-ciphertext pairs encrypted by the related keys.

Additionally, the cryptanalysis methods can be classified according to the usefulness of the cryptanalysis results, such as total break, global deduction, local deduction, information deduction, etc.

- Total break denotes that the attacker recovers the private key.
- Global deduction means that the attacker obtains only a functionally equivalent encryption/decryption algorithm but not the private key.
- Local deduction denotes that the attacker explores additional plaintexts or ciphertexts not previously known.
- Information deduction means that the attacker gets some Shannon information [21] about plaintexts or ciphertexts not previously known.

The recovered information decreases in the order listed. For example, the key is recovered in total break while not in global deduction, the additional plaintexts or ciphertexts are recovered in local deduction while not in information deduction.

3.3 Multimedia Compression

Multimedia data is often of large volumes, which is compressed before storage or transmission in order to save space or bandwidth cost. For text or binary data, various compression methods have been reported, such as Huffman coding [22], arithmetic coding [23], Golomb coding [24], run-length coding [25], LZW coding [26], etc. They are constructed based on entropy theory [27]. Taking Huffman coding, for example, the data sample that happens with high frequency is assigned a short codeword, while the data sample that seldom happens is assigned a long codeword. The encoded data stream can be decoded without loss. In contrast, in image, audio or video, there is much more redundancy, which can be compressed by entropy-based coding methods.

3.3.1 Audio Compression

Generic compression algorithms for text or binary data are not suitable for audio data, for two reasons. First, audio data is composed of the samples in time order, which has quite different properties compared with text or binary data. Second, audio data should be used for real-time applications. Typical codecs have been designed for audio data, such as DPCM [28], LPC [29], CELP [30], mp3 [31], AAC [32], etc. According to the operation domain, they can be classified into two types, time domain codec and frequency domain codec.

In time domain, the audio segment can be predicted with the adjacent segment, which is often used to reduce the bit rate of the compressed audio signal. Typical compression methods include Differential Pulse-Code Modulation (DPCM), Linear Predictive Coding (LPC) and Code Excited Linear Prediction (CELP). DPCM is a lossless codec that encodes the Pulse-code Modulation (PCM) values as differences between the current and the previous value. DPCM has been extended and used in the standard wideband speech codecs, G.721 [33] and G.723 [34]. The LPC uses a sound generator model to whiten the audio signal before quantization. The linear prediction can reduce the data volumes to be encoded. The decoding process uses the linear predictor to shape the quantization noise into the decoded signal's spectrum. The CELP uses the source-filter model of speech production through linear prediction, and uses an adaptive codebook and a fixed codebook. Compared with LPC, CELP obtains the audio signal with much better quality. It has been extended to various codecs and used widely for speech coding.

Frequency domain methods compress audio signal greatly by making use of the human psychoacoustic model. Generally, audio signal is partitioned into segments, and the segments are encoded in order. First, the audio segment is transformed by modified discrete cosine transform (MDCT) [31, 32] from time domain in frequency domain. The transformation is also called analysis filterbank. Second, the psychoacoustic model corresponding to the audio segment is computed. Third, the parameters of the psychoacoustic model are used to control the quantization of the frequency coefficients. The psychoacoustic model not only considers the audibility of the audio content but also considers the bit rate of the compressed signal. The typical codec includes MPEG1 Layer 3 (mp3) and Advanced Audio Coding (AAC). Figure 3.4 shows the general architectures of encoder and decoder.

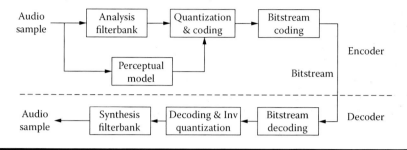

Figure 3.4   General architectures of the encoder and decoder in the frequency domain.

3.3.2 *Image Compression*

Various methods for image compression have been reported and widely used. According to the loss property, they can be classified into two types, lossless compression and lossy compression. In lossless compression, the encoded image can be recovered without loss. Most of the typical lossless compression methods are constructed based on entropy coding. For example, PCX image is encoded with run-length coding [25], and GIF and TIFF images are encoded with the adaptive dictionary algorithm LZW [26].

Compared with lossless compression, lossy compression causes quality degradation to the decoded image. The advantage is that it can obtain a higher compression ratio. Such encoding methods as fractal coding [35] and vector quantization [36] have been used in previous decades because of their high compression ratio. Fractal coding is based on the assumption that there is self-similarity among an image. The vector quantization encodes an image segment by segment through designing a vector table. However, because natural images have various properties, these coding methods cannot get the decoded image with high quality. Generally, they are often used together with some other codecs.

Recently, compression methods based on transformation are used more widely. These methods adopt the sensitivity of the human eyes to reduce the image data's bit rate. First, the image is converted from RGB color space to YCbCr space, and the image data in Cb or Cr space is downsampled, because human eyes are more sensitive to Y space than to Cb or Cr space. Second, the data is transformed from space domain to frequency domain, which concentrates the energy on low frequency. As shown in Figure 3.5, in encoding, the original image is transformed into frequency domain, then quantized, and finally followed by entropy encoding. In decoding, the image stream is decoded by entropy codec, then inversely quantized, and finally inversely transformed. Generally, the transformation and quantization operations cause losses to image quality. The typical transformation includes discrete cosine transformation (DCT) and wavelet. The entropy coding may be

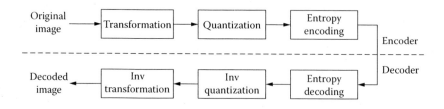

Figure 3.5   General architectures of image encoder and decoder based on transformation.

run-length coding, Huffman coding or arithmetic coding. The typical codec based on DCT is JPEG [37], which is the most widely used image compression standard. The codecs based on wavelet include Set Partitioning in Hierarchical Trees (SPIHT) [38] and JPEG2000 [39]. Here, JPEG2000 is the latest image compression standard. Generally, in DCT-based codecs, the image is partitioned into 8×8 blocks, each block is transformed by 8×8 DCT, and the blocks are encoded one by one. In wavelet-based codecs, the whole image is transformed by a wavelet transformation, then the transformed image is partitioned into blocks, and the blocks are encoded one by one.

3.3.3 Video Compression

Video is often composed of a continuous image sequence. According to this property, video compression is constructed based on image compression. The simplest method is to encode each image one by one, for example, MJPEG [40]. The apparent disadvantage is that the relation between adjacent frames is not used. Some video coding standards have been published that combine transform coding and motion coding. Here, transform coding denotes the transformation-based coding for images, whereas motion coding means to encode the frame by referencing the adjacent frames. The standards include H.261 [41], MPEG-1 [42], MPEG-2 [43], H.263 [44], MPEG-4 part 2 [45] and MPEG-4 AVC/H.264 [46]. Among them, H.261 was the first video compression standard used in videoconferencing and videotelephony. MPEG-1 is used for Video CD (VCD) while the video quality is low, MPEG-2 is used for DVD and has been widely used in digital video broadcasting and cable distribution systems. H.263 is the improved version of h.261 and is used for videoconferencing, videotelephony, and Internet video. MPEG-4 part 2 improves the quality of MPEG-2 and H.263 and has been used for Internet, broadcast, mobile phone, etc., and MPEG-4 AVC/H.264 has wide compression capability and is gaining a wide variety of applications.

For the coded based on both transform coding and motion coding, the typical method is to partition the video into Groups of Pictures (GOPs), classify the images in GOP into three kinds, i.e., I-frame, P-frame and B-frame, and encode each frame

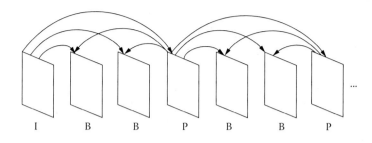

Figure 3.6   General architecture of a GOP.

with the image coding methods. As shown in Figure 3.6, the I-frame is encoded independently of the method for image coding, for example, JPEG; the P-frame is encoded by referencing the previous I-frame or P-frame; and the B-frame is encoded by referencing the adjacent P-frame or I-frame.

Taking MPEG2 for example, the GOP encoding and decoding processes are shown in Figure 3.7. In I-frame encoding, each block is encoded with DCT transformation, quantization (Q), run-length encoding (RLE), and variable-length

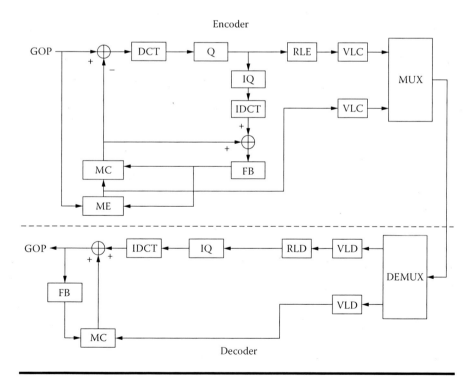

Figure 3.7   General architecture of GOP encoding and decoding.

coding (VLC). In I-frame decoding, each block is decoded by variable-length decoding (VLD), run-length decoding (RLD), inverse quantization (IQ), and inverse DCT (IDCT). For the P- or B-frame, each block is encoded by motion estimation (ME) and motion compensation (MC) besides DCT, Q, RLE, and VLC. The frame buffer (FB) stores the previous frames that are used as the references. The decoding of P- or B-frames is symmetric to the encoding process.

3.3.4 Scalable Coding

Scalable coding, also called progressive coding, generally refers to generating content with reduced quality by directly manipulating the bitstream without decompression or recompression. It is especially useful for previewing images while downloading them (e.g., in a web browser) or for providing variable quality access to bandwidth-limited channels. For example, a TV program can be transmitted from a cable transmission system to a mobile channel by cutting the bitstream adaptively. Generally, there are several types of scalability.

- Temporal scalability: the content can be downsampled in time axis by directly dropping some parts from the bitstream.
- Spatial scalability: the content's spatial resolution can be reduced by directly cutting some parts from the bitstream.
- Quality scalability: the content's quality can be reduced by directly removing some parts in the bitstream.

Some existing codecs have these scalabilities. For example, an MPEG-2 codec can encode video content into progressive layers that correspond to different quality. SPIHT encodes an image into the progressive bitstream that satisfies quality scalability. JPEG2000 encodes an image into the progressive bitstream according to spatial scalability and quality scalability. MPEG4 Fine Granularity Scalability (FGS) [47] encodes a video into progressive layers with quality scalability and temporal scalability. Scalable Video Coding (SVC) [48], constructing on MPEG4 AVC/H.264, produces a scalable bitstream with all the temporal scalability, spatial scalability and quality scalability.

Additionally, the scalability may be measured by the granularity. For MPEG-2 codec, the media data is encoded into two layers, the base layer and the enhancement layer. However, the enhancement layer is of large scalable granularity and can only be cut completely. As shown in Figure 3.8(a), if only a part of the enhancement layer is cut, the remaining enhancement part cannot be decoded. For MPEG4 FGS, the media data is encoded into two layers, a base layer and an enhancement layer. In contrast, the enhancement layer is of fine scalable granularity and can be cut arbitrarily. As shown in Figure 3.8(b), if only a part of the enhancement layer is cut, the remaining enhancement part can still be decoded. For such codecs as SPIHT or SVC, the media data is encoded into the scalable bitstream with

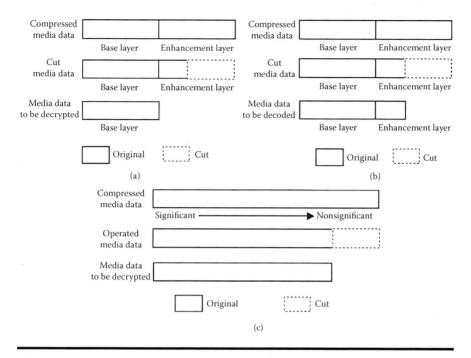

Figure 3.8 The scalable granularity of different scalable codecs.

fine scalable granularity. As shown in Figure 3.8(c), the cutting operation does not affect the correct decoding.

3.4 Multimedia Communication

Multimedia communication solves the problem of transmitting multimedia content from the sender to the receiver via various channels. Generally, the method depends on the transmission networks and the application scenarios.

3.4.1 Transmission Network

Multimedia content can be transmitted over different networks using different methods, such as unicasting [49], broadcasting [50], multicasting [51], or p2p [52], as shown in Figure 3.9.

Unicasting denotes sending the content to a single destination. Because only the sender and receiver join in the transmission, this transmission method can be used to send personal information that is sensitive, such as the decryption key or access right.

Broadcasting means to distribute content that can be received by every receiver in the network. In this scenario, the sender needs only to send the content once,

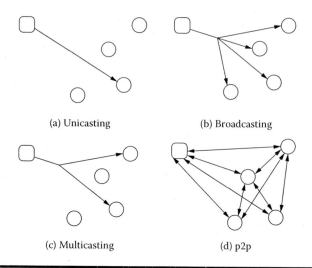

(a) Unicasting   (b) Broadcasting

(c) Multicasting   (d) p2p

Figure 3.9   Different multimedia transmission methods.

and all the receivers can receive it. According to this property, the transmission method is often used in services with a large number of customers, such as audio broadcasting or TV broadcasting. The typical broadcasting network includes satellite radio, satellite TV, cable TV, Internet, etc.

Multicasting is the trade-off between unicasting and broadcasting. It means to transmit the content to a group of receivers simultaneously. It needs only to transmit the content once, and all the receivers in the group can receive the content. The typical application is for IP Multicasting that implements the multicasting functionality on the IP routing level. The receiver requests the content by providing the router's address, and the sender delivers the content to the router corresponding to the IP group address. This transmission method is suitable for video-on-demand services in local networks.

Peer-to-peer (p2p) is different from the traditional server-client model. For example, in such server-client models as FTP [53] file transfer, the clients initiate the download/upload request, and the FTP server reacts to and satisfies the request. In a p2p network, each peer acts as both "clients" and "servers" of the other peers. This transmission method exploits the diverse connectivity and the cumulative bandwidth of the network peers compared with traditional modes. Thus, it is suitable for such applications as file sharing, telephony, IPTV, etc.

3.4.2 Application Scenarios

Different application scenarios need different transmission methods. The typical application includes live TV [54], video-on-demand (VOD) [55], file downloading [56], streaming media [57], etc.

Live TV refers to television programming broadcast in real time. It is used especially for real-time news. Because the source content is produced in real time, the transmission process should be as efficient as possible in order to keep the content "live." Typical Live TV systems include satellite systems, cable systems and mobile TV systems.

Video-on-demand (VOD) provides customers the ability to select and watch a video or clip over a network in an interactive manner. Here, the interactivity includes the functionality of pause, fast forward, fast rewind, slow forward, slow rewind, jump to previous or future frame, etc. Different from live TV, the content is often recorded and stored at the sender side beforehand.

Videoconferencing allows two or more locations to interact via two-way video and audio transmissions simultaneously. It has also been called visual collaboration. In each location, the following devices are required: video input/output, audio input/output, and data transmission channel. Thus, the core technologies are audio/video compression and real-time transmission. Generally, compression rates of up to 1:500 can be achieved, and the transmission network may be ISDN [58] or IP.

Streaming is a technique to provide real-time multimedia services. In these services, the multimedia content can be decoded and displayed at the receiver side during content delivery. Thus, it is different from such file transfer protocols as FTP that needs to download the whole content before displaying it. The Real-time Streaming Protocol (RTSP) [59], Real-time Transport Protocol (RTP) [60], and the Real-time Transport Control Protocol (RTCP) [61] were specifically designed to stream media over networks. A media stream can be on demand or live. On demand streams are stored on a server for a long period of time, and are available to be transmitted at a user's request. Live streams are only available at one particular time, as in a video stream of a live sporting event. Streaming media is widely used in p2p or multicasting network.

Downloading means to download the media content from the sender and store it at the receiver side, which is different from streaming. It is used in the scenario when the media content needs to be stored or recorded. The typical application is FTP or HTTP [62] that needs to download all the content before being able to edit or display it. Real-time operation is important to streaming, but not required by downloading.

3.5 Digital Watermarking

Digital watermarking [4] is the technique of embedding some information into multimedia content by modifying the media content slightly. The information, called a watermark, can be extracted from the marked media content. Generally, a watermarking algorithm has three properties [63], imperceptibility, robustness, and security. Imperceptibility means that there is no perceptual difference between the marked media and the original media. Robustness denotes the watermark's ability

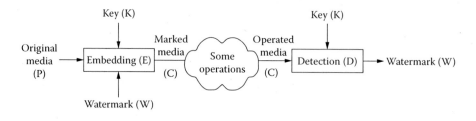

Figure 3.10 Architecture of a watermarking system.

to survive some operations on the marked media, such as compression, adding noise, filtering, etc. The security is determined by the ability to resist an unauthorized watermark extractor.

A typical watermarking system is shown in Figure 3.10. The watermark embedding process is defined as

$$C = E(P, W, K).$$

Here, K is the key used to control the embedding process. For example, K is used to encrypt the watermark W before embedding W into P, or K is used to select suitable embedding positions from P. The marked media content C is modified into C' by some operations, and then, the watermark extraction process is

$$W = D(C', K).$$

Digital watermarking can be used in various applications, such as ownership authentication, content authentication, secret communication, etc. Additionally, a watermark can be used to trace illegal distributors. For example, when the media content is transmitted to the customers, the customer's identification code is embedded into the content. Thus, each customer obtains a unique copy containing his ID code. As shown in Figure 3.11, the embedding process is defined as

$$C_i = E(P, W_i, K).$$

Here, W_i is the ith customer's ID code, C_i is the ith customer's media copy, and $i = 0, 1, \ldots, n - 1$. If a copy is redistributed to unauthorized customers, such as over the Internet, the ID code can be extracted from the copy and identify the illegal distributor. Taking the ith copy for example, the customer code can be detected by

$$W_i = D(C_i, K).$$

In this scenario, because the customer code acts as the watermark and can be used to identify the customer, the technique is also named digital fingerprinting [64]. It is often used to design secure multimedia distribution schemes together with encryption techniques [65].

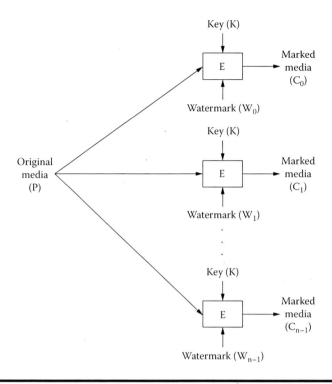

Figure 3.11 Architecture of a fingerprinting system.

3.6 Summary

The chapter introduces the basic techniques of multimedia content encryption, including cryptography, multimedia compression, multimedia communication, and digital watermarking. The concepts, terms and architectures presented in this chapter will provide the basic for the following chapters.

References

[1] R. A. Mollin. 2006. *An Introduction to Cryptography.* Boca Raton, FL: CRC Press.
[2] K. Sayood. 2005. *Introduction to Data Compression,* 3rd ed. San Francisco, CA: Morgan Kaufmann.
[3] K. Rao, Z. Bojkovic, and D. Milovanovic. 2006. *Introduction to Multimedia Communications: Applications, Middleware, Networking.* NY: Wiley-Interscience.
[4] I. J. Cox, M. L. Miller, and J. A. Bloom. 2002. *Digital Watermarking.* San Francisco, CA: Morgan-Kaufmann.
[5] FIPS 46-2. Data Encryption Standard. November 1993.
[6] FIPS 197. Advanced Encryption Standard (AES). November 2001.

[7] X. Lai, and J. L. Massey. 1991. A Proposal for a New Block Encryption Standard, Springer-Verlag, Lecture Notes on Computer Science (LNCS), Vol. 473: 389–404.

[8] RSA Labs Public Key Cryptography Standards (PKCS). November 1993. IEEE P1363.

[9] ANSI X9.62. Elliptic Curve Cryptography (ECC). 1999.

[10] T. ElGamal. 1985. A public-key cryptosystem and a signature scheme based on discrete logarithms. *IEEE Transactions on Information Theory* IT-31(4): 469–472.

[11] National Institute of Standards and Technology (NIST). 2004. *Recommendation for the Triple Data Encryption Algorithm (TDEA) Block Cipher.* Special Publication 800-67. Gaithersburg, MD: NIST.

[12] M. Luby. 1996. *Pseudorandomness and Cryptographic Applications.* Princeton, NJ: Princeton University Press.

[13] National Institute of Standards and Technology (NIST). 2006. *Recommendation for Random Number Generation Using Deterministic Random Bit Generators.* Gaithersburg, MD: NIST.

[14] S. Lian, J. Sun, and Z. Wang. 2007. A chaotic stream cipher and the usage in video encryption. *International Journal of Chaos, Solitons and Fractals* 34(3): 851–859.

[15] B. Schneier. 1996. Section 17.1 RC4. In *Applied Cryptography*, 2nd ed. NY: John Wiley & Sons.

[16] P. Ekdahl, and T. Johansson. 2003. A new version of the stream cipher SNOW. 9th Annual International Workshop, SAC 2002, St. John's, Newfoundland, Canada, August 15–16, 2002. Lecture Notes in Computer Science, 2595, 47–61.

[17] M. Boesgaard, M. Vesterager, T. Pedersen, J. Christiansen, and O. Scavenius. 2003. Rabbit: A high-performance stream cipher. Proceedings FSE 2003. Lecture Notes in Computer Science, 2887, 307–329.

[18] National Institute of Standards and Technology (NIST). 2004. *Recommendation for Block Cipher Modes of Operation.* Gaithersburg, MD: NIST.

[19] W. F. Friedman. 1993. *Military Cryptanalysis: Transposition and Fractionating Systems.* Walnut Creek, CA: Aegean Park Press.

[20] E. Biham. 1994. New types of cryptanalytic attacks using related keys. *Journal of Cryptology* 7(4): 229–246.

[21] C. Shannon. 1949. Communication theory of secrecy systems. *Bell System Technical Journal* 28: 656–715.

[22] D. A. Huffman. 1952 (September). A method for the construction of minimum-redundancy codes. *Proceedings of the* Institute of Radio Engineers, 40(9): 1098–1101.

[23] G. Glen, and J.R. Langdon. 1984. An introduction to arithmetic coding. *IBM Journal of Research and Development* 28(2): 135–149.

[24] S. W. Golomb. 1966. Run-length encodings. *IEEE Transactions on Information Theory* 12(3): 399–401.

[25] Run-Length Encoding (RLE). http://en.wikipedia.org/wiki/Run-length_encoding

[26] T. A. Welch. 1984. A technique for high-performance data compression. *Computer* 17: 8–19.

[27] T. Cover, and J. Thomas. 1991. *Elements of Information Theory.* NY: John Wiley & Sons.

[28] W. N. Waggener. 1999. *Pulse Code Modulation Systems Design*, 1st ed., Boston, MA: Artech House.

[29] B. S. Atal. 2006. The history of linear prediction. *IEEE Signal Processing Magazine* 23(2): 154–161.

[30] M. R. Schroeder, and B. S. Atal. 1985. Code-excited linear prediction (CELP): High-quality speech at very low bit rates. In *Proceedings of the IEEE International Conference on Acoustics, Speech, and Signal Processing (ICASSP)*, Vol. 10, 937–940.

[31] ISO/IEC International Standard IS 11172-3. Information Technology-Coding of Moving Pictures and Associated Audio for Digital Storage Media at up to about 1.5 Mbits/s-Part 3: Audio.

[32] MPEG4 Part3, ISO/IEC 14496-3: Advanced Audio Coding.

[33] G.723: Extensions of ITU Recommendation G.721 adaptive differential pulse code modulation to 24 and 40 kbit/s for digital circuit multiplication equipment application.

[34] G.726: ITU Recommendation for 40, 32, 24, 16 kbit/s Adaptive Differential Pulse Code Modulation (ADPCM).

[35] V. Drakopoulos, P. Bouboulis, and S. Theodoridis. 2006. Image compression using affine fractal interpolation on rectangular lattices. *Fractals* 14(4): 259–269.

[36] T.-S. Chen, and C.-C. Chang. 1997. A new image coding algorithm using variable-rate side-match finite-state vector quantization. *IEEE Transactions on Image Processing* 6(8): 1185–1187.

[37] W. B. Pennebaker, and J. L. Mitchell. 1993. *JPEG Still Image Compression Standard*. New York: Van Nostrand Reinhold.

[38] A. Said. and W. A. Pearlman. 1996. A new fast and efficient image codec based on set partitioning in hierarchical trees. *IEEE Transactions on Circuits and Systems for Video Technology* 6(3): 243–250.

[39] ISO/IECFCD15444-1: Information technology - JPEG2000 image coding system - Part 1: Core coding system, March 2000.

[40] ISO/IECFCD15444-3: Information technology - JPEG2000 image coding system - Part 3: Motion JPEG2000, 2002.

[41] ITU Recommendation, H.261 (03/93): Video codec for audiovisual services at p ×64 kbit/s.

[42] Coding of moving pictures and associated audio. In Committee Draft Standard ISO11172: 1/2/3. 1991.

[43] ISO/MPEG-2. ISO 13818-2: Coding of moving pictures and associated audio, 1994.

[44] ITU-T Recommendation H.263-Video Coding for Low Bit Rate Communication.

[45] MPEG-4 part 2 (ISO/IEC 14496-2): Advanced Simple Profile (ASP).

[46] H.264/MPEG4 Part 10 (ISO/IEC 14496-10): Advanced Video Coding (AVC), ITU-T H.264 standard.

[47] W. Li. 2001. Overview of fine granularity scalability in MPEG-4 Video Standard. *IEEE Transactions on Circuits and Systems for Video Technology* 11(3): 301–317.

[48] H. Schwarz, D. Marpe, and T. Wiegand. 2007. Overview of the Scalable Video Coding Extension of the H.264/AVC Standard. *IEEE Transactions on Circuits and Systems for Video Technology* 17(9): 1103–1120.

[49] Unicast, http://en.wikipedia.org/wiki/Unicast

[50] Broadcasting, http://en.wikipedia.org/wiki/Broadcasting_%28networks%29

[51] X. Li, M. Ammar, and S. Paul. 1999. Video multicast over the Internet. *IEEE Network Magazine* 13(2): 46–60.

[52] S. Androutsellis-Theotokis, and D. Spinellis. 2004. A survey of peer-to-peer content distribution technologies. *ACM Computing Surveys* 36(4): 335–371.

[53] RFC 3659 — Extensions to FTP (File Transfer Protocol). P. Hethmon. March 2007.

[54] Live TV. http://en.wikipedia.org/wiki/Live_TV

[55] Video on Demand (VOD), http://en.wikipedia.org/wiki/Video_on_demand

[56] File Transfer, http://en.wikipedia.org/wiki/File_transfer

[57] Streaming Media, http://en.wikipedia.org/wiki/Streaming_media

[58] (Integrated Services Digital Network) ISDN, http://www.itu.int

[59] RTSP, RFC 2326, Real Time Streaming Protocol (RTSP).

[60] RTP, RFC 1889, Obsolete, RTP: A Transport Protocol for Real-Time Applications.

[61] RTCP (Real time control protocol), RFC 3550.

[62] HTTP, Latest release of HTTP/1.1 specification is dated June 1999, RFC 2616.

[63] M. Barni, and F. Bartolini. 2004. *Watermark Systems Engineering*. NY: Marcel Dekker.

[64] M. Wu, W. Trappe, Z. J. Wang, and K. J. R. Liu. 2004. Collusion-resistant fingerprinting for multimedia. *IEEE Signal Processing Magazine* 21(2): 15–27.

[65] S. Lian, Z. Liu, Z. Ren, and H. Wang. 2006. Secure distribution scheme for compressed video stream. In *Proceedings 2006 IEEE International Conference on Image Processing (ICIP2006)* 1953–1956.

Chapter 4

Complete Encryption

4.1 Definition of Complete Encryption

Complete encryption is the algorithm that encrypts multimedia content completed without considering format. As shown in Figure 4.1(a), this kind of algorithm encrypts raw data or compressed data directly with traditional or novel ciphers under the control of an encryption key. The decryption process is symmetric to the encryption process, as shown in Figure 4.1(b). Here, the symmetric cipher is preferred by considering of media data's large volumes and asymmetric cipher's high computational complexity. The encryption system's security and efficiency are determined by the adopted cipher, while its effect on compression ratio and format compliance depends on the relation between encryption operation and compression operation.

4.2 Classification of Complete Encryption

Some algorithms have been proposed to encrypt media data completely, which can be classified into various types according to their properties.

First, according to the relation between encryption and compression, they can be classified into two types, raw data encryption and compressed data encryption. In raw data encryption, the media data is encrypted before it is compressed. Since the encryption operation changes the performance of the multimedia data, for example, the adjacent relation of image pixels, the compression ratio may be changed greatly. In compressed data encryption, the media data is first compressed and then encrypted. Because the compression operation reduces the data volume of original

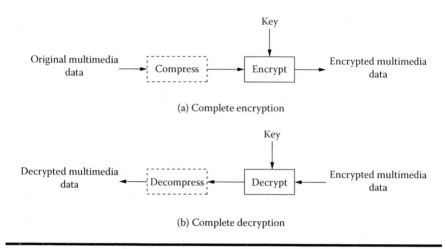

(a) Complete encryption

(b) Complete decryption

Figure 4.1   General architecture of complete encryption.

media data, the followed encryption operation saves much computational cost. They will be presented in detail in Section 4.3.

Second, according to the properties of the encryption algorithm, they can be classified into permutation algorithm, random modulation algorithm, confusion-diffusion algorithm, and partial DES algorithm, etc. These will be introduced in Section 4.4.

Third, according to the implementation, they can be classified into software-based algorithms and hardware-based algorithms. Software-based algorithms are easily updated, whereas hardware-based algorithms can provide high security and computational efficiency. Thus, there is a trade-off between security and efficiency, which will be presented in Section 4.5.

4.3 Encryption Algorithms for Raw Data or Compressed Data

4.3.1 Raw Data Encryption

Raw data encryption algorithms encrypt raw data directly, which does not consider the compression process. Generally two kinds of method are often used, permutation algorithms and confusion-diffusion algorithms.

Permutation algorithms permute the uncompressed multimedia data with the control of an encryption key. For example, pay-per-view TV programming [1] is scrambled by a permutation algorithm based on pseudorandom numbers. This algorithm permutes the TV field line by line, and assigns each field a unique subkey. Alternatively, Pseudorandom Space Filling Curves are used to change the scanning order of the image or video pixels [2], which reduces the intelligibility of the image or video. In the algorithm, the scanning order is controlled by an encryption key that determines the shape of the space filling curves. Additionally, chaotic maps are used to permute the images or videos [3, 4]. Taking the coordinates of the image or video pixels as the initial value of a chaotic map, the coordinates are changed by iterating the chaotic map. The control parameters of the chaotic map act as the encryption or decryption key. Furthermore, some mathematical transforms are used to permute image or audio data, such as the magic transform [5] or Fibonacci transform [6], which change the pixels' position in the order controlled by the key. These permutation algorithms are of low computing or power cost, and often make the permuted images unintelligible. However, they are not secure enough from a cryptographic viewpoint; for example, they are not secure against select-plaintext attacks [7]. They will be analyzed in detail in the next section.

Confusion-diffusion algorithms not only permute the pixels of multimedia data but also change the amplitudes of the pixels. For example, block ciphers based on chaotic maps [8–10] first permute the pixel's position and then change the pixel's amplitude by diffusing the changes from one pixel to another adjacent one. These algorithms are often of higher security than permutation algorithms, but of lower efficiency than permutation algorithms.

In general, these raw data encryption algorithms decrease the understandability of multimedia data through changing the adjacent relation among image pixels. They are of low cost in terms of computing or power. However, they often change the statistic properties of multimedia data, which will affect the following compression operation. Therefore, these algorithms are often used to encrypt multimedia data that do not need to be compressed, such as TV signals, BMP images, audio broadcasting, etc.

4.3.2 Compressed Data Encryption

A compressed data encryption algorithm, as shown in Figure 4.1, encrypts the compressed data stream directly. The existing algorithms can be classified into two types: data stream permutation algorithms and data stream encryption algorithms.

Data stream permutation algorithms permute the compressed data stream directly. For example, Qiao and Nahrstedt [11] proposed that an MPEG1/2 video stream can be permuted with a byte as a unit, for three reasons: first, it is convenient, second, a byte is meaningless, and third, it is random. In this algorithm, the video stream is often partitioned into segments and then permuted segment by segment. The segment size determines the security level. That is, the bigger the segment is, the higher the security. This kind of algorithm is of low cost, while also being of low security. Thus, it is more suitable for applications that require real-time operation with low security.

Data stream encryption algorithms encrypt the compressed data stream with modified or novel ciphers. For example, Qiao and Nahrstedt [12] proposed VEA

(Video Encryption Algorithm) that encrypts only the even half of plaintext with DES and obtains the odd half through XOR-operation on the even half and the odd half. Thus, the encryption time is reduced to nearly 46% compared with DES. Tosun and Feng [13] extended the algorithm by partitioning the even half into two halves again, and decreased the encryption time to nearly a quarter. Agi and Gong [14] proposed a video encryption algorithm that uses the key stream produced by DES algorithm to modulate the video stream bit by bit. Romeo et al. [15] proposed an RPK algorithm that combines a stream cipher with a block cipher. Wee and Apostolopoulos [16, 17] proposed an algorithm that encrypts the progressive data stream with a stream cipher layer by layer. These algorithms are often of high security and benefit from the traditional ciphers adopted. However, they change the file format, which makes the encrypted data inoperable without decryption. Additionally, because all the compressed data are encrypted, the efficiency is not very high.

4.4 Typical Encryption Algorithms

4.4.1 Permutation

Permutation is the one of the oldest encryption algorithms. It transforms the plaintext into an unintelligible form by changing the adjacent relationship of the pixels. Taking random pixel permutation, random line permutation and chaos-based permutation for examples, we present their implementation in detail.

4.4.1.1 Random Pixel Permutation

Random pixel permutation changes the position of plaintext pixels under the control of a random sequence. Let $P = p_0 p_1 \ldots p_{N-1}$, $C = c_0 c_1 \ldots c_{N-1}$ and $S = s_0 s_1 \ldots s_{N-1}$ ($0 \leq s_i < 1$, $i = 0, 1, \ldots, N-1$) be the plaintext, ciphertext, and random sequence, respectively. Then, the example of the random pixel permutation process is denoted by the following pseudo program.

```
Initialize C=P;
Temp=c0;
For i=1:N
  j=floor(si*i);
  Temp=cj;
  cj=ci;
  ci=Temp;
end
```

Here, floor(x) denotes the biggest integer no bigger than x. As can be seen, the permutation operation only changes the position of the plaintext pixel, while keeping the pixels amplitude unchanged, which can be seen in Figure 4.2(a). According to

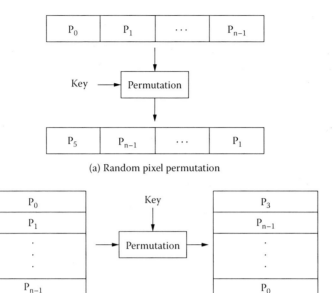

(a) Random pixel permutation

(b) Random line permutation

Figure 4.2 Examples of random pixel/line permutation.

the implementation, this encryption method has low computational cost, and can be used to encrypt arbitrary digital content. The pixel may be a bit in a string, a sample in an audio sequence or a pixel in an image.

4.4.1.2 Random Line Permutation

Similar to random pixel permutation, random line permutation changes the position of the plaintext lines under the control of a random sequence. Set $P = p_0 p_1 \ldots p_{N-1}$, $C = c_0 c_1 \ldots c_{N-1}$ and $S = s_0 s_1 \ldots s_{N-1}$ $(0 \le s_i < 1, i = 0, 1, \ldots, N-1)$ as the plaintext, ciphertext, and random sequence, respectively. Here, p_i or c_i denotes the ith line in the plaintext or ciphertext. Then, a simple random line permutation process can be denoted by a similar pseudo program:

```
Initialize C=P;
Temp=c₀;
For i=1:N
 j=floor(sᵢ*i);
 Temp=cⱼ;
 cⱼ=cᵢ;
 cᵢ=Temp;
end
```

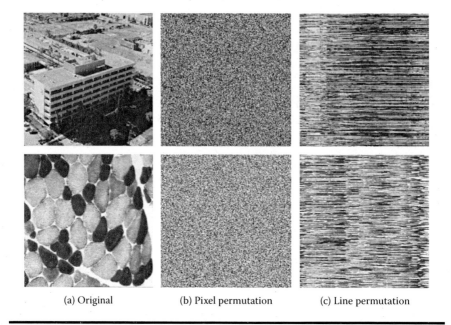

<table>
<tr><td>(a) Original</td><td>(b) Pixel permutation</td><td>(c) Line permutation</td></tr>
</table>

Figure 4.3 Examples of image permutation.

Here, floor(x) denotes the biggest integer no bigger than x. Thus, the permutation operation only changes the position of a plaintext line, while keeping the pixel's amplitude unchanged, which can be seen in Figure 4.2(b). Similar to pixel permutation, this encryption algorithm is also of high efficiency. It can be used for not only digital content, but also for analog media. For example, it has been used to encrypt European TV programs by permuting the TV lines [1]. Figure 4.3 shows the image encrypted by random pixel permutation or random line permutation. As can be seen, the encrypted images are too chaotic to be understood, which maintains high perceptual security.

4.4.1.3 Chaotic Map-Based Permutation

A chaotic map is a dynamic system that can be easily denoted by mathematical equations [8–10]. Generally, for a chaotic map, the initial value acts as its input, the control parameters determine its action, and the output is the sequence produced by the iterated maps, as shown in Figure 4.4. It has some typical properties suitable for data encryption. Given the initial value, the chaotic map generates the sequence with random properties. The sequence is sensitive to the initial value, that is, two initial values with a slight difference will cause great differences after multiple iterations. Additionally, the sequence is very sensitive to the control parameters, that is, different parameters produce different sequences.

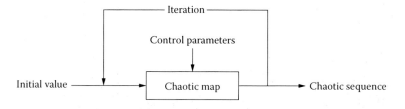

Figure 4.4   General architecture of a chaotic map.

As shown in Figure 4.4, a chaotic map is similar to a cipher. The initial value of the chaotic map is regarded as the plaintext, the control parameters as the key and the output sequence as the ciphertext. A chaotic map-based permutation uses the chaotic map to permute the plaintext. Generally, the original position and permutated position act as the plaintext and ciphertext, respectively; the area preserving the chaotic map acts as the cipher, and the chaotic map's parameters act as the key. Area-preserving chaotic maps includes Cat map, Baker map, and Standard map, etc. For example, in the unit square area, the Cat map shown in Figure 4.5 is defined as

$$\begin{pmatrix} x_{k+1} \\ y_{k+1} \end{pmatrix} = C(x_k, y_k) = \begin{pmatrix} 1 & b \\ a & ab+1 \end{pmatrix} \begin{pmatrix} x_k \\ y_k \end{pmatrix} \bmod 1,$$

where $0 \leq x_k, y_k < 1$, k is the iteration time, and a, b are the control parameters satisfying $0 \leq a, b < 1$. This map is reversible. Similarly, the Baker map shown in Figure 4.6 is defined as

$$(x_{k+1}, y_{k+1}) = B(x_k, y_k) = \begin{cases} (2x_k, y_k/2), & 0 \leq x_k < 1/2 \\ (2x_k - 1, y_k/2 + 1/2), & 1/2 \leq x_k \leq 1 \end{cases},$$

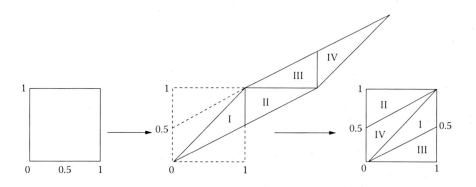

Figure 4.5   A Cat map.

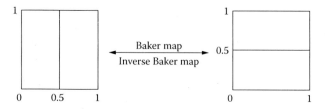

Figure 4.6   A Baker map.

where $0 \le x_k, y_k < 1$, and k is the iteration time. This map is reversible. A standard map is defined as

$$\begin{bmatrix} x_{k+1} \\ y_{k+1} \end{bmatrix} = S(x_k, y_k) = \begin{bmatrix} x_k + y_k \bmod 2\pi \\ y_k - q\sin(x_k + y_k) \bmod 2\pi \end{bmatrix},$$

where $0 \le x_k, y_k < 1$, k is the iteration time, and q is the control parameter that satisfies $q > 0$. This map is reversible.

These chaotic maps should be discretized before being used for permutation. Taking a Cat map, for example, in a square area $M \times M$, the discrete Cat map is defined as

$$\begin{pmatrix} x_{k+1} \\ y_{k+1} \end{pmatrix} = C(x_k, y_k) = \begin{pmatrix} 1 & b \\ a & ab+1 \end{pmatrix}\begin{pmatrix} x_k \\ y_k \end{pmatrix} \bmod M,$$

where (x_k, y_k) is the position of the pixel in the original square area and satisfies $0 \le x_k, y_k < M$, k is the iteration time, a, b are the control parameters satisfying $0 \le a, b < M$ (a and b are integers), and (x_{k+1}, y_{k+1}) is the position of the pixel in the permuted square area. Thus, a and b make up the key controlling the permutation process. The inverse permutation process is defined as

$$\begin{pmatrix} x_k \\ y_k \end{pmatrix} = C^{-1}(x_{k+1}, y_{k+1}) = \begin{pmatrix} 1 & b \\ a & ab+1 \end{pmatrix}^{-1}\begin{pmatrix} x_{k+1} \\ y_{k+1} \end{pmatrix} \bmod M,$$

where $C^{-1}()$ is the inverse function of $C()$.

According to the chaotic map's property, the pixel in the original square area can be moved to another position far from its original position with the rise of iteration time. Taking a 256 × 256-sized image, for example, the discrete Cat map with $a = 1$ and $b = 2$ is used to permute it. The image permuted by seven iterations of the chaotic map is more confused than one permuted by one chaotic map, as shown in

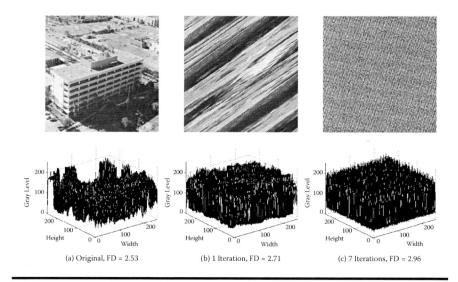

(a) Original, FD = 2.53   (b) 1 Iteration, FD = 2.71   (c) 7 Iterations, FD = 2.96

Figure 4.7   Images encrypted by different numbers of chaotic map iterations.

Figure 4.7. Furthermore, the fractal dimensions of different images also prove it. Thus, more iteration times should be used in order to improve the perceptual security.

Permutation algorithms are often of high encryption efficiency. The encrypted content is often unintelligible. However, the permutation operation is fragile to known-plaintext attacks and select-plaintext attacks. In these attacks, the comparison between plaintext and ciphertext is done to find the permutation principle. Thus, to strengthen these algorithms, variable key encryption should be adopted. That is, to encrypt different plaintexts with different keys.

4.4.2   Random Modulation

Random modulation is one of the simplest methods [18] of encrypting media content with sequential properties, such as audio sequences. In an audio sequence, the audio samples are processed one by one, and an efficient operation is needed to meet real-time applications. Random modulation changes the audio sequence with a random sequence by making use of simple operations, such as bitwise XOR or module addition. Among them, XOR is suitable for digital content, while module addition is more suitable for analog content. As shown in Figure 4.8, the plaintext $P = p_0 p_1 \ldots p_{N-1}$ is modulated into the ciphertext $C = c_0 c_1 \ldots c_{N-1}$ by the random sequence $R = r_0 r_1 \ldots r_{N-1}$. If the media content is in digital form, then p_i, c_i, r_i all are in bit forms. And thus, the encryption operation and decryption operation are defined as

$$\begin{cases} c_i = p_i \oplus k_i \\ p_i = c_i \oplus k_i \end{cases}.$$

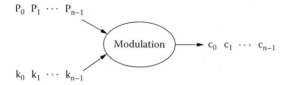

Figure 4.8   Architecture of random modulation.

Otherwise, if the media content is analog, then p_i, c_i, r_i all are in floating data. And thus, the encryption operation and decryption operation are defined as

$$\begin{cases} c_i = (p_i + k_i) \bmod 1 \\ p_i = (c_i - k_i) \bmod 1 \end{cases}.$$

Taking the audio sequence in analog form, for example, the sequence is encrypted by random modulation. Generally, the encrypted audio sequence is completely unintelligible, as shown in Figure 4.9. Thus, this kind of encryption algorithm is often of high perceptual security. Additionally, the encryption/decryption is time efficient. However, for random modulation, its cryptographic security depends on

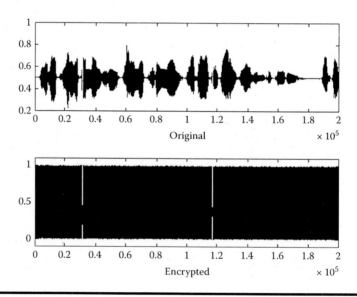

Figure 4.9   Results of random modulation-based audio sequence encryption (peak signal-to-noise ratio = 8.67 dB).

the random sequence generator. A good random sequence generator that produces a long sequence with randomness can confirm the security of the cipher. Otherwise, the cipher may be fragile to known-plaintext or select-plaintext attacks. However, to date, no means have been developed to generate pure random sequences, and most of the existing generators only produce pseudorandom sequences, which cannot confirm the security of the cipher.

4.4.3 Confusion-Diffusion Algorithm

A confusion-diffusion algorithm encrypts the plaintext by n ($n > 0$) iterations, each of which is composed of a confusion operation and a diffusion operation. This encryption mode was first reported by Shannon [19], and has been used in designing such strong block ciphers as DES, IDEA, and AES. These traditional ciphers (DEA, IDEA, AES, etc.) often work on plaintext with small size (typically, 64 bits or 128 bits). Considering that multimedia data is often of large volume and high redundancy, the ciphers supporting large plaintext size may bring some good properties, such as high encryption efficiency and perceptual security. First, the encryption operation may work pixel by pixel or byte by byte instead of bit by bit, which may improve the computing efficiency. Second, plaintext with larger size is dealt with together, which can reduce the redundancy between different plaintext blocks.

One of the typical confusion-diffusion algorithms is constructed based on a chaotic map [8–10]. As shown in Figure 4.10, in encryption, the plaintext is first permuted by a chaotic map, then diffused by a diffusion function, and the two

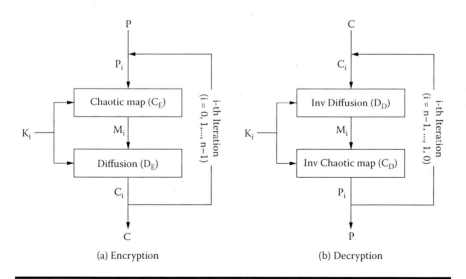

(a) Encryption (b) Decryption

Figure 4.10 **Architecture of the chaos-based confusion-diffusion algorithm.**

operations are repeated for n times. For the ith iteration, let P_i, M_i and C_i be the plain-block, middle-block, and cipher-block, respectively. Then, the encryption operations in the ith iteration are defined as

$$\begin{cases} M_i = C_E(P_i, K_i) \\ C_i = D_E(M_i, K_i) \end{cases}.$$

Here, the ith key K_i controls the ith permutation C_E and diffusion D_E operations. The encryption operations are repeated from 0 to the $n-1$th iteration.

The decryption is symmetric to encryption: the ciphertext is first inversely diffused by the diffusion function, and then inversely permuted by the chaotic map, and the two operations are repeated for n times. For each iteration, the decryption operations are defined as

$$\begin{cases} M_i = D_D(C_i, K_i) \\ P_i = C_D(M_i, K_i) \end{cases}.$$

Here, the ith key K_i controls the ith inverse permutation C_D and inverse diffusion D_D operations. The decryption operations are repeated from the $n-1$th iteration to 0.

The chaos-based permutation operation or inverse permutation operation has been defined in Section 4.4.1.3. The size of the square area can be varied. For example, if the image size is 256×256, then the square area of the chaotic map is 256×256.

The diffusion function spreads changes in one pixel to other pixels. The typical diffusion function is based on XOR or exponential. Let M_i be the ith pixel in the original data, Q_{i-1} and Q_i the $i-1$th and ith pixel in the diffused data, and L the pixels gray level. For the one based on XOR, the diffusion function is

$$Q_i = M_i \oplus Q_{i-1},$$

where Q_{-1} is controlled by the key. The inverse diffusion function is

$$M_i = Q_i \oplus Q_{i-1}.$$

Similarly, for the one based on exponential, the diffusion function is

$$Q_i = \left(M_i + Q_{i-1}^2\right) \bmod L,$$

where Q_{-1} is controlled by the key. And, the inverse diffusion function is

$$M_i = \left(Q_i - Q_{i-1}^2\right) \bmod L.$$

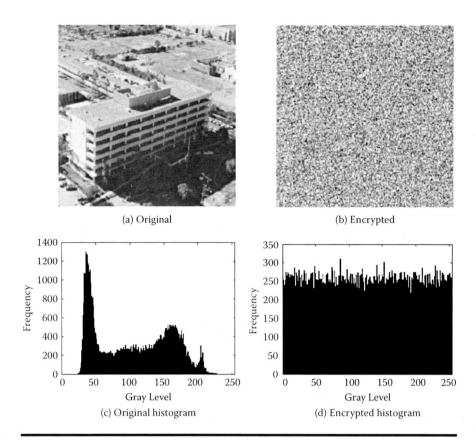

(a) Original

(b) Encrypted

(c) Original histogram

(d) Encrypted histogram

Figure 4.11   Results of the encrypted image.

By selecting suitable parameters for the chaotic map and iteration time, the cipher can obtain good performances in cryptographic aspect and perceptual aspect. Generally, the preferred plaintext size is no smaller than 64×64. If a standard map is used, the iteration time n should be no smaller than 4, if a Cat map is used, the iteration time should be no smaller than 6, and if a Baker map is used, the iteration time should be no smaller than 12. Taking $N \times N = 256 \times 256$, $n = 6$, Cat map and exponential based diffusion function for example, the encrypted image and the corresponding histogram are shown in Figure 4.11. The key sensitivity and plaintext sensitivity corresponding to different values of n are shown in Figure 4.12. For key sensitivity, the KS corresponding to every key bit (total 64 bits) is tested, and for plaintext sensitivity, the PS corresponding to every plain-images pixel (total 256×256 pixels) is tested. As can be seen, the encrypted image is unintelligible, its histogram is near in uniform distribution, the key sensitivity is near to 50% when n is not smaller than 3, and the plaintext sensitivity is near to 50% when n is not smaller than 6. These properties prove the cipher performs well in cryptographic

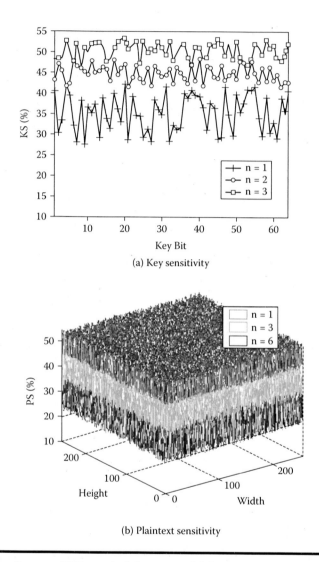

(a) Key sensitivity

(b) Plaintext sensitivity

Figure 4.12   Key sensitivity and plaintext sensitivity.

security and perceptual security. The encryption efficiency of the chaos-based cipher is compared with that of 3DES. Generally, the bigger the N is, the more efficient the chaos-based cipher is. When N is no smaller than 128, the encryption time ratio between chaos-based cipher and 3DES is no bigger than 10%.

4.4.4   Partial DES

As has been mentioned above, such traditional ciphers as DES, 3DES, or AES are not suitable for media data encryption. They should be modified in order to obtain

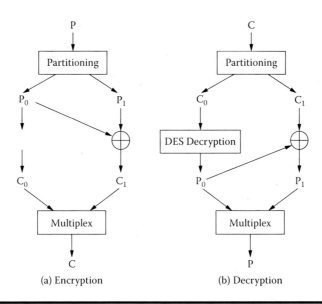

Figure 4.13 Architecture of partial AES encryption/decryption.

high perceptual security and encryption efficiency. The well-known modification is the partial DES algorithm [12, 13]. This scheme encrypts only parts of media content with the strong block cipher, while encrypting the other parts by such simple operations as XOR.

Figure 4.13 shows a simple example that encrypts only half of the media data. First, original media data P is partitioned into two parts with the same length, P_0 and P_1. Then, the two parts are encrypted into two cipher-parts, C_0 and C_1, according to the following method.

$$\begin{cases} C_0 = E(P_0, K) \\ C_1 = P_0 \oplus P_1 \end{cases}.$$

Finally, C_0 and C_1 are combined together with the same manner in partitioning, which produces the ciphertext C.

The decryption process is symmetric to the encryption process. First, the ciphertext C is partitioned into two parts, C_0 and C_1, in the same manner. Then, the two parts are decrypted into P_0 and P_1 according to the following method.

$$\begin{cases} P_0 = D(C_0, K) \\ P_1 = P_0 \oplus C_1 \end{cases}.$$

Finally, P_0 and P_1 are combined together with the same manner in partitioning, which produces the plaintext P.

With this method, the encryption time is reduced to 46% compared with the original complete encryption. The partial DES method can be extended from partitioning two parts to partitioning 4, 8, …, 2^{-v} parts. Thus, the encryption time can be reduced to nearly 2^{-v} compared with the original complete encryption.

4.5 Security and Efficiency

Complete encryption is also termed format-independent encryption because it regards multimedia data as binary data and encrypts multimedia data without considering the file format. Generally, there is a trade-off between security and encryption efficiency. The algorithms with higher security are often of higher computing complexity. For example, traditional strong ciphers [20], such as DES, IDEA, AES, RSA, etc., encrypt text or binary data directly without considering the file format. These ciphers have been included in the protocols IPSec and SSL (Secure Socket Layer), and the package CryptoAPI. However, they are often of high computational cost, and not suitable for real-time multimedia applications or power-limited terminals. Alternatively, the random modulation methods [18] are often based on stream ciphers that are implemented by simple operations. Thus, they are efficient in implementation. However, their security cannot be confirmed completely. To obtain the trade-off, some means are proposed, among which hardware implementation and lightweight algorithm are two typical solutions.

4.5.1 Hardware Implementation

To improve the encryption algorithms' efficiency, hardware implementation is a suitable solution. Secure processing architectures are proposed in [21], which includes an embedded processor, a cryptographic hardware accelerator, and a programmable security protocol engine. For the core encryption algorithms, some experiments are done to show their suitability. For example, hardware implementation of 3DES [22] can obtain small area and reasonable throughput even though 3DES turns out to be very source-consuming. It is suitable for some applications in WLAN (Wireless Local Area Network). Compared with such block ciphers as 3DES, stream ciphers have some good properties, such as immunity to error propagation, increased flexibility, and greater efficiency. Some improved stream ciphers are implemented and tested in hardware [23] such as WEP (Wired Equivalent Privacy), IWEP (Improved Wired Equivalent Privacy) and RC4 (Ron's Cipher #4) [24]. These hardware-based stream ciphers have high efficiency and are suitable for real-time multimedia communication even in a wireless/mobile environment. As can be seen, hardware implementation improves the cipher's computing efficiency, but it also creates some problems, for example, the high cost to update the algorithms.

4.5.2 Lightweight Encryption

Compared with hardware implementation, software implementation is cheaper and more flexible for updates. For wireless applications, some lightweight encryption algorithms have been proposed, such as WEP, IWEP, RC2, RC4, RC5, etc. The software efficiency of RC2 and RC4 is tested in [25]. It is shown that software implementation of these ciphers can meet the requirements of such wireless applications as multimedia e-mail, multimedia notes, telephone-quality audio, video conferencing or MPEG video interaction, etc. The disadvantage is that their security is limited by the efficient operation, and their performance is limited by the computer system configuration.

4.6 Performance Comparison

As mentioned above, various complete encryption algorithms have been reported. these algorithms are compared and listed in Table 4.1. According to their performances in security, compression efficiency, encryption efficiency, format compliance, and application suitability. For all the algorithms, if they are used to encrypt media data before compression, then the subsequent compression will be greatly affected, and thus, the compression ratio change belongs to CL2. Otherwise, if they are used to encrypt compressed media data, then they do not change the data size, and thus, the compression ratio change belongs to CL0. Among these algorithms, permutation and random modulation have lower security compared with other algorithms, confusion-diffusion and partial DES cost more time and power than other algorithms, and hardware implementation and lightweight algorithms provide a better trade-off between the various performances. According to their performances, confusion-diffusion is suitable for multimedia storage, partial DES is suitable for multimedia transmission, permutation or hardware implementation is more suitable for real-time interaction, and random modulation or lightweight algorithm is more suitable for wireless/mobile communication.

4.7 Summary

In this chapter, some complete encryption algorithms are introduced, including random permutation, random modulation, confusion-diffusion, partial DES, hardware implementation, and lightweight algorithms. By analyzing their performances in security, compression ratio, encryption efficiency, and format compliance, the

Table 4.1 Performance Comparison of Various Complete Encryption Algorithms

Encryption Algorithm	Security	Compression Ratio	Encryption Efficiency	Format Compliance	Application Suitability
Permutation	SL2	CL0 or CL2	EL0	FL0	Real-time interaction
Random modulation	SL2	CL0 or CL2	EL0	FL0	Wireless/mobile communication
Confusion-diffusion	SL0	CL0 or CL2	EL1	FL0	Multimedia storage
Partial DES	SL0	CL0 or CL2	EL1	FL0	Multimedia transmission
Hardware implementation	SL0	CL0 or CL2	EL0	FL0	Real-time interaction
Lightweight algorithm	SL0	CL0 or CL2	EL0	FL0	Wireless/mobile communication

suitable application scenarios are determined. This chapter is expected to provide valuable information to researchers.

References

[1] European Commission for Electrotechnical Standardization (CENELEC). 1992 (December). Access control system for the MAC/packet family: EUROCRYPT. European Standard EN 50094. Brussels: CENELEC.

[2] Y. Matias, and A. Shamir. 1987. A video scrambling technique based on space filling curves. *Proceedings of Advances in Cryptology-CRYPTO'87*. Lecture Notes in Computer Science, 293, 398–417.

[3] J. Scharinger. 1998. Kolmogorov systems: Internal time, irreversibility and cryptographic applications. In *Proceedings of the AIP Conference on Computing Anticipatory Systems*, vol. 437, ed. D. Dubois. Woodbury, NY: American Instute of Physics.

[4] F. Pichler, and J. Scharinger. 1996. *Finite Dimensional Generalized Baker Dynamical Systems for Cryptographic Applications*. Lecture Notes in Computer Science, 1030, 465–476.

[5] Ye Yongwei, Yang Qinghua, and Wang Yingyu. 2003. Magic cube encryption for digital image using chaotic sequence, *Journal of Zhejiang University of Technology* 31(2): 173–176.

[6] D. Qi, J. Zou, and X. Han. 2000. A new class of scrambling transformation and its application in the image information covering. *Science in China Series E* (English Edition) 43(3): 304–312.

[7] Shujun Li, Chengqing Li, Guanrong Chen, and Nikolaos G. Bourbakis. 2004. A General Cryptanalysis of Permutation-Only Multimedia Encryption Algorithms. IACRs Cryptology ePrint Archive: Report 2004/374'.

[8] J. Fridrich. 1997. Image encryption based on chaotic maps. *Proceedings of IEEE Nonlinear Signal and Image Processing Workshop* 1105–1120.

[9] G. R. Chen, Y. B. Mao, and C. K. Chui. 2004. A symmetric image encryption scheme based on 3D chaotic cat maps. *Chaos, Solitons and Fractals* 12: 749–761.

[10] Y. B. Mao, G. R. Chen, and S. G. Lian. 2004. A novel fast image encryption scheme based on the 3D chaotic Baker map. *International Journal of Bifurcation and Chaos* 14(10): 3613–3624.

[11] L. Qiao, and K. Nahrstedt. 1998. Comparison of MPEG encryption algorithm. *International Journal on Computers and Graphics* 22(4): 437–448.

[12] L. Qao, and K. Nahrstedt. 1997. A new algorithm for MPEG video encryption. In *Proceeding of the First International Conference on Imaging Science, Systems and Technology (CISST'97)*, Las Vegas, NV, July, 21–29.

[13] A. S. Tosun, and W. C. Feng. 2001. Lightweight security mechanisms for wireless video transmission. *Proceedings International Conference on Information Technology: Coding and Computing*, April 2–4, 157–161.

[14] I. Agi, and L. Gong. 1996. An empirical study of MPEG video transmissions. In *Proceedings of the Internet Society Symposium on Network and Distributed System Security*, San Diego, CA, February 137–144.

[15] A. Romeo, G. Romdotti, M. Mattavelli, and D. Mlynek. 1999. Cryptosystem architectures for very high throughput multimedia encryption: The RPK solution. *Proceedings of 6th IEEE International Conference on Electronics, Circuits and Systems, ICECS '99*, vol. 1, September 5–8, 261–264.

[16] S. J. Wee, and J. G. Apostolopoulos. 2001. Secure scalable video streaming for wireless networks. In *Proceedings of the IEEE International Conference on Acoustics, Speech, and Signal Processing*, vol. 4, Salt Lake City, UT, May, 2049–2052.

[17] S. J. Wee, and J. G. Apostolopoulos. 2001. Secure scalable streaming enabling transcoding without decryption. In *Proceedings of the IEEE International Conference on Image Processing*, vol. 1, Thessaloniki, Greece, October 7–10, 437–440.

[18] Chih-Hsu Yen, Yu-Shiang Lin, and Bing-Fei Wu. 2007. An efficient implementation of a low-complexity MP3 algorithm with a stream cipher. *Multimedia Tools and Applications* 35(3): 335–355.

[19] C. Shannon. 1949. Communication theory of secrecy systems. *Bell System Technical Journal* 28: 656–715.

[20] Richard A. Mollin. 2006. *An Introduction to Cryptography*. Boca Raton, FL: CRC Press.

[21] A. Raghunathan, S. Ravi, S. Hattangady, and J.-J. Quisquater. 2003. Securing mobile appliances: New challenges for the system fesigner. *2003 Europe Conference and Exibition in Design, Automation and Test*, 176–181.

[22] P. Hamalainen, M. Hannikainen, T. Hamalainen, et al. 2001. Configurable hardware implementation of triple DES encryption algorithm for wireless local area network. In *Proceedings IEEE International Conference on Acoustics, Speech and Signal Processing*, 1221–1224.

[23] J. Goodman, and A. P. Chandrakasan. 1998. Low power scalable encryption for wireless systems. *Wireless Networks* 4: 55–70.

[24] K. Tikkanen, M. Hannikainen, T. Hamalainen, and J. Saarinen. 2000. Hardware implementation of the improved WEP and RC4 encryption algorithms for wireless terminals. In *Proceedings European Signal Processing Conference*, September, 2289–2292.

[25] A. Ganz, S. H. Park, and Z. Ganz.1998. Inline network encryption for multimedia wireless LANs. *Proceedings IEEE Military Communications Conference*, October.

Chapter 5

Partial Encryption

5.1 Definition of Partial Encryption

Partial encryption is the algorithm that encrypts only a part of the multimedia content while leaving other parts unchanged. As shown in Figure 5.1(a), multimedia data is first partitioned into two parts, then, one part is encrypted by traditional or novel ciphers under the control of the key, and the other part is not changed, and finally, the two parts are combined together. The decryption process is symmetric to the encryption process, as shown in Figure 5.1(b). In practice, media content is composed of various parts. Let the original media content P be partitioned into n parts, that is, $P_0, P_1, \ldots, P_{N-1}$. Then, in partial encryption, only n $(0 < n \leq N)$ parts are encrypted according to

$$C_i = E_i(P_i, K_i), \quad i = 0, 1, \ldots, n-1,$$

where C_i, K_i, $E_i()$ are the ith $(i = 0, 1, \ldots, n-1)$ cipher-part, key, and encryption algorithm, respectively. Typically, different parts can be encrypted with the same key or same encryption algorithm. Thus, the encrypted media content C is composed of $C_0, C_1, \ldots, C_{n-1}$ and $P_n, P_{n+1}, \ldots, P_{N-1}$. Similarly, in decryption, only n parts are decrypted according to

$$P_i = D_i(C_i, K_i), \quad i = 0, 1, \ldots, n-1,$$

where K_i and $D_i()$ are the ith $(i = 0, 1, \ldots, n-1)$ key and decryption algorithm, respectively. Typically, different parts can be decrypted with the same key or same decryption algorithm. Thus, the decrypted media content P is composed of $P_0, P_1, \ldots,$ and P_{N-1}.

As can be seen, a partial encryption scheme aims to improve the encryption efficiency by reducing the volume of encrypted data. However, the security is often

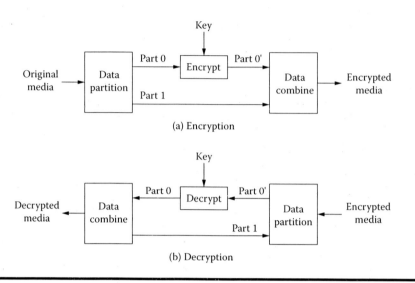

(a) Encryption

(b) Decryption

Figure 5.1   Architecture of partial encryption/decryption.

an open issue in this scheme. The efficiency and security depend on two key steps, that is, data partitioning and part selection.

5.1.1   Data Partitioning

The media data can be partitioned according to various aspects, for example, the coding parameters, the object-background, the blocks, or the encoding layers. Thus, the "part" may denote the parameter, object, block or coding layer, etc.

Considering that media data may be encoded into the data stream composed of various coding parameters, the data stream can be partitioned into different parts, and each part corresponds to a coding parameter, as shown in Figure 5.2(a). Then, the partial encryption is carried out by encrypting only some important parameters. For example, in video coding based on MPEG2 [1], the encoded data stream is composed of such parameters as syntax information, DCT coefficients and motion vector.

An encryption scheme with an object or background as a "part" is shown in Figure 5.2(b), where the media content, e.g., image or video, can be partitioned into objects and background. Intuitively, this partitioning may work in the spatial domain. For example, in such codecs as JPEG2000 [2] or MPEG4 [3], the region-of-interest coding or object-based coding provide the convenience to encrypt only the important objects. As an example, only the background is encrypted, while other objects are left unchanged.

An encryption scheme with a data block as a "part" is shown in Figure 5.2(c), which is suitable for block-based encoding, such as JPEG [4], MPEG2 or H.263

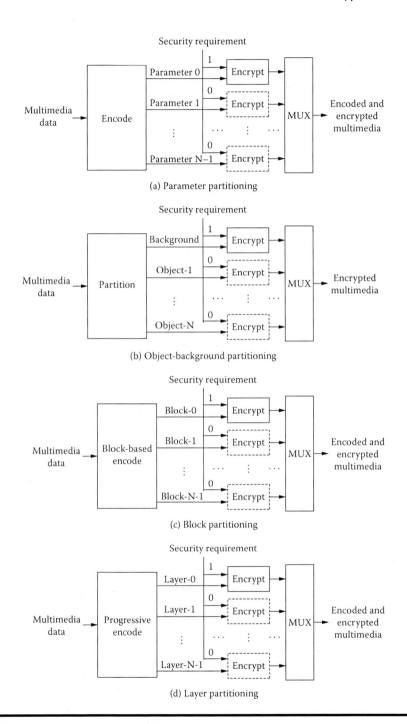

Figure 5.2 Various data partitioning in partial encryption.

[5]. In these codecs, the media data is partitioned into data blocks, and the blocks are encoded one by one. In partial encryption, only some blocks are encrypted with a cipher, while the other blocks are left unchanged. This encryption scheme can be combined with these media compression codecs.

An encryption scheme with a data layer as a "part" is shown in Figure 5.2(d), which is suitable for progressive encoding, such as EZW [6], SPIHT [7], JPEG2000, MPEG4, or SVC [8]. These encoding processes encode multimedia data into progressive data streams. By encrypting some of the layers, multimedia data can be encrypted into unintelligible forms. For example, only the base-layer stream is encrypted, while the others are left unencrypted.

5.1.2 Part Selection

Part selection means to select suitable data parts from the partitioned ones, which will be encrypted by a cipher. Generally, part selection should consider two performances, that is, encryption efficiency and security. For efficiency, the fewer the data parts that are encrypted, the higher the encryption efficiency that can be obtained. For security, the more the data parts are encrypted, the higher the system's security is. Thus, the trade-off between efficiency and security should be considered.

Considering only the security, there are some principles for selecting the data parts.

- First, the encrypted data part should be independent from the unencrypted part. It avoids having to deduce the encrypted part from the unencrypted part. Thus, for the attacker, knowing the unencrypted part provides little help in recovering the encrypted part.
- Second, the encrypted data part should be significant to human perception. Intuitively, the data part that is sensitive to media content's intelligibility is preferred to be encrypted. Thus, encrypting the sensitive data part makes the media content unintelligible.
- Third, the data part should be encrypted by a strong cipher with high security. Although a strong cipher has high computational complexity and requires much time, the system's encryption efficiency can be maintained by reducing the number of encrypted parts.

5.2 Classification of Partial Encryption

Various partial encryption algorithms have been reported, which can be classified into different types according to different properties. According to the type of multimedia data, existing partial encryption algorithms can be classified into partial audio encryption, partial image encryption, and partial video encryption. According to the relation between encryption and compression, the algorithms can be classified into partial encryption for raw data, partial encryption during compression, and partial encryption for compressed data. According to the encryption operation, the algorithms can be classified into permutation-based encryption, stream cipher-based encryption and block cipher-based encryption. The typical algorithms for image, audio, or video are introduced and analyzed in detail below.

5.3 Partial Image Encryption

For image data, the partial encryption algorithms can be classified into three types according to the relation between the encryption operation and the compression operation. These are raw image encryption, partial image encryption during compression, and partial encryption of the compressed image.

5.3.1 Raw Image Encryption

For the raw image, the data partitioning can be done with respect to bit-planes or objects. In the former case, the image is partitioned into bit-planes, that is, from the most significant one to the least significant one. Thus, L most significant bit-planes are encrypted, while the other $N - L$ ones are left unencrypted. The architecture of the partial encryption scheme is shown in Figure 5.3. Taking the image with 8 bit-planes (8-bit gray level) for example, the image encryption algorithm obtains high security when more than 4 bit-planes are encrypted, as shown in Figure 5.4. Here, the bit-planes are encrypted with AES cipher. However, the selection of the number of encrypted bit-planes varies with image content [9], which makes it difficult to give a common number that is suitable for all natural images.

In the second case, the image is partitioned into objects and background [10]. Here, the objects denote the meaningful things in the image, including the person's face, car, animal, plane, etc., while the background denotes the things that are not noted, for example, the sky, river, highway, etc. Thus, the partial encryption scheme can encrypt only the objects in the image. Seen from the example shown

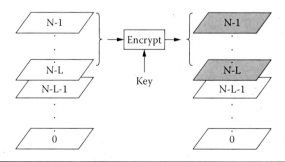

Figure 5.3   Architecture of bit-plane encryption.

(a) Original   (b) Encrypted

Figure 5.4   Bit-plane encryption of raw images.

in Figure 5.5, only the person's face is encrypted in an image, and thus, the image has no commercial value.

5.3.2 Partial Image Encryption during Compression

During image compression, some parameters can be encrypted before being processed by the subsequent operations. These methods reduce the volume of the encrypted data and improve the encryption efficiency. However, since the encryption

(a) Original   (b) Encrypted

Figure 5.5   Object encryption of raw images.

operation changes the properties of the parameters, the subsequent operations may cause great changes to the compression ratio. Partial encryption algorithms during run-length coding and wavelet-based encoding are introduced below.

5.3.2.1 Partial Encryption in Run-Length Coding

Run-length coding is used for PCX images [11]; it compresses the image by making use of the relationship between adjacent pixels. For an image, each of its lines is encoded with run-length coding that produces the (Level,Run) pairs. Here, Level denotes the pixel's gray level, and Run denotes the number of adjacent pixels that have the same Level. The partial encryption scheme encrypts only the Run in each (Level,Run) pair.

Suppose each image line is composed of N pixels, and it is encoded into the (Level,Run) pairs, that is, (x_0, p_0), (x_1, p_1), ..., (x_{n-1}, p_{n-1}). Here, $0 \leq p_i < 64$ ($i = 0$, 1, ..., $n - 1$), and $p_0 + p_1 + ... + p_{n-1} = N$. Then, the encryption process shown in Figure 5.6 is defined as

$$c_i = \mathrm{E}(p_i, K).$$

Here, $E()$ is a cipher, K is the key and c_i is the encrypted run. Then, the encrypted data stream is composed of the pairs, that is, (x_0, c_0), (x_1, c_1), ..., (x_{n-1}, c_{n-1}). Here, $0 \leq c_i < 64$ ($i = 0, 1, ..., n - 1$). The decryption process is symmetric to the encryption process. That is, the (x_i, c_i) pairs are decrypted by

$$p_i = \mathrm{D}(c_i, K).$$

Here, $D()$ is the decryption operation symmetric to $E()$.

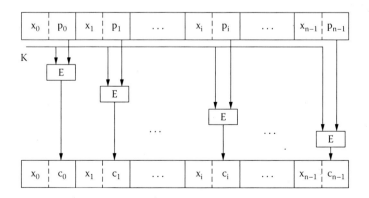

Figure 5.6 Partial encryption in run-length coding.

(a) Original

(b) Encrypted

Figure 5.7   Images encrypted by the run-length encryption algorithm.

Because this encryption algorithm changes the Runs in the (Level,Run) pairs, the adjacent relation between image pixels is greatly changed, which makes the encrypted image unintelligible. Figure 5.7 shows some images encrypted by this algorithm. Here, the stream cipher, RC4, is used as the functions $E()$ and $D()$. Note that, if the (Level,Run) pairs are further encoded with a subsequent entropy coding, the compression ratio will be changed greatly. That is the algorithm's main disadvantage.

5.3.2.2   Partial Encryption in Wavelet-Based Codec

Wavelet transformation is often used in image coding, which has such properties as multiresolution composition. Typical codecs are EZW, SPIHT and JPEG2000.

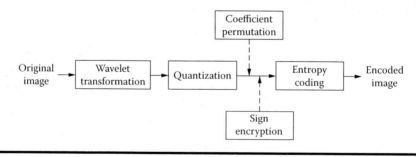

Figure 5.8   Partial encryption in wavelet-based codec.

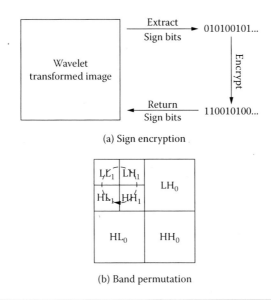

(a) Sign encryption

(b) Band permutation

Figure 5.9 Sign encryption and band permutation of wavelet coefficients.

Generally, the image is first transformed by discrete wavelet transformation, then quantized, and finally encoded with entropy coding. Encryption operations, such as sign encryption and coefficient permutation, are often inserted between quantization and entropy coding, as shown in Figure 5.8.

Sign encryption means to encrypt the sign bits of the wavelet coefficients with a cipher, as shown in Figure 5.9(a). If the coefficient is smaller than 0, then its sign bit is denoted by 0, otherwise, by 1. Thus, the sign-bits can be extracted from the coefficients, encrypted by traditional ciphers and returned to the corresponding coefficients. In the existing wavelet-based codecs, the sign bits are often encoded independently, and thus, sign encryption does not change the compression ratio. Considering that, after quantization, most of the coefficients in high frequency become zeros, the sign bits to be encrypted will be greatly reduced, and thus, the security will be decreased.

Coefficient permutation means to permute the coefficients in the frequency bands with existing permutation methods. The coefficients' positions are permuted in each frequency band or subblock. Here, the frequency band may be partitioned into subblocks. For example, Uehara [12] proposed an algorithm that permutes wavelet coefficients in different frequency bands. Zeng and Lei [13] proposed an algorithm that permutes wavelet coefficients in each subblock. Lian, Sun, and Wang [14] proposed an algorithm based on SPIHT codec, which permutes the coefficients in a tree-based structure. Taking two-level wavelet transformation for example, the transformed image is composed of seven frequency bands, that is, LL_1, LH_1,

HL_1, HH_1, LH_0, HL_0, and HH_0. Some of the frequency bands in low frequency can be permuted, which degrades the quality of the image greatly. As shown in Figure 5.9(b), the first four frequency bands (LL_1, LH_1, HL_1, and HH_1) are permuted, while the other three are left unchanged. Because coefficient permutation changes the adjacent relation between coefficients, band permutation causes great

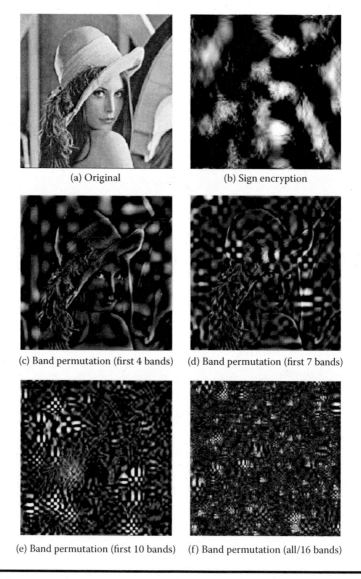

(a) Original   (b) Sign encryption

(c) Band permutation (first 4 bands)   (d) Band permutation (first 7 bands)

(e) Band permutation (first 10 bands)   (f) Band permutation (all/16 bands)

Figure 5.10 Images encrypted in wavelet-based codec.

changes to the compression ratio. Additionally, the permutation itself is not secure enough against known-plaintext or select-plaintext attacks.

The images encrypted by sign encryption and band permutation are shown in Figure 5.10. Here, five-level wavelet transformation is tested, the wavelet coefficients are permuted in the first B ($B = 4, 7, 10, 16$) bands, (b) is the one encrypted by sign encryption, (c) is the one encrypted by permuting the first four frequency bands, and (d) is the one encrypted by permuting all the frequency bands. As can be seen, only the images encrypted by sign encryption and all band permutations are unintelligible, while the other permutation that does not permute all the bands produces a still intelligible image. Thus, the partial band permutation in wavelet domain is not secure in perception.

5.3.3 Partial Encryption of the Compressed Image

Another image encryption method is encrypting the parameters of the compressed image selectively. This method is widely used for JPEG2000 image. For example, Wee and Apostolopoulos [15] proposed a secure scalable streaming scheme for Motion-JPEG2000 codec, which encrypts scalable data with AES or 3DES and supports some direct operations. Pommer and Uhl [16] proposed a selective encryption scheme for wavelet-packet encoded images, which encrypts only some wavelet trees and is of low cost. Norcen and Uhl [17] proposed a selective encryption scheme for JPEG2000 bitstream, which encrypts 20% of the compressed bitstream except format information.

In JPEG2000 codec, images are transformed into different frequency bands that represent different fidelity or resolution. Additionally, each subband is partitioned into a number of code blocks, and each code block is encoded bit-plane by bit-plane from the most significant one to the least significant one. In addition, each bit-plane is encoded with three passes, among which, the significant pass encodes some significant coefficients' signs, the refinement pass encodes some coefficients' bit value, and the cleanup pass encodes both some significant coefficients' signs and coefficients' position information. Thus, the compressed data stream can be partitioned into three layers, that is, subband, bit-plane, and code pass. The relationships between the data layers are shown in Figure 5.11. A partial encryption scheme that encrypts some suitable data layers has been proposed by Lian et al. [18]. In this scheme, the subbands, bit-planes and code passes can be encrypted selectively. For example, only the four subbands in low frequency are selected, for each subband, only the four most important bit-planes are selected, and for each bit-plane, only the significant pass is encrypted. Figure 5.12 shows the encrypted results corresponding to Peppers (128×128, colorful, five-level wavelet transform) and Plane (512×512, colorful, six-level wavelet transform), where the AES cipher is adopted. As can be seen, this algorithm is secure in perception. Additionally, it does not change the compression ratio and supports the direct operation of cutting some data layers directly.

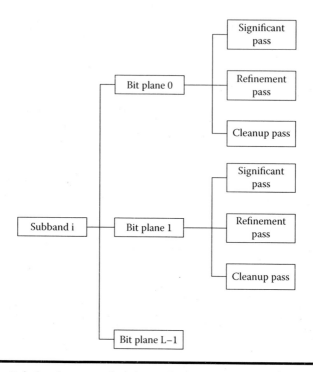

Figure 5.11 Relation between data layers in JPEG2000 image.

5.4 Partial Audio Encryption

Audio data is often encoded before being transmitted in order to save transmission bandwidth. The existing partial audio encryption algorithms encrypt audio data during or after audio compression. The former one encrypts some parameters during audio encoding. For example, the speech data is encrypted by encrypting only the parameters of fast Fourier transformation during the speech encoding process [19]. This kind of algorithm degrades audio quality by changing the transformed parameters, which is time efficient. However, the parameter changes often affect the subsequent entropy coding, and thus, changes the compression ratio. Another method encrypts some sensitive parameters in the compressed data stream. For example, an algorithm based on G.729 [20, 21] is proposed to encrypt telephone-bandwidth speech. This algorithm partitions the bitstream into two classes, the most perceptually relevant one and one other. The perceptually relevant bitstream is encrypted while the other one is left. It is reported that encrypting about 45% of the bitstream achieves content protection equivalent to full encryption. For another

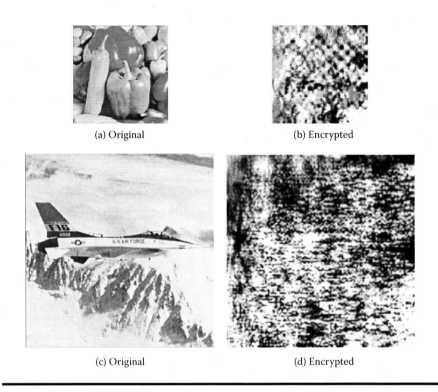

(a) Original   (b) Encrypted

(c) Original   (d) Encrypted

Figure 5.12   Results of partial JPEG2000 image encryption.

example, in mp3 codec, various parameters are produced, including the synchronization information, Huffman codes, scale factors, bit allocation information, etc. The partial encryption scheme [22, 23] only encrypts some of them with a cipher, for example, bit allocation information or some Huffman codes. Figure 5.13 shows an example for mp3 audio encryption. In decryption, the right parameters are recovered, which are used to decode the audio data. For encrypting only a few parameters, this kind of encryption algorithm is often of high efficiency.

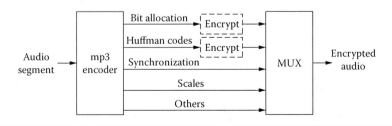

Figure 5.13   Architecture of mp3 audio encryption.

5.5 Partial Video Encryption

Compared with image or audio data, video data is often of larger volume, and needs to be compressed in order to reduce the transmission bandwidth. Additionally, the compressed video data stream is composed of many parameters, such as format information, texture information and motion information. Taking MPEG1/2/4 for example, the format information includes sequence header/end, GOP header, picture header, etc., the texture information is composed of Differential DC and AC, and the motion information denotes motion vector difference (MVD). In MPEG4 AVC/H.264 [24], the texture information includes intraprediction mode besides DC and AC, and the motion information includes interprediction mode besides MVD. Generally, video encryption algorithms can be classified into two types, encryption after compression and encryption during compression.

5.5.1 Encryption after Compression

This kind of algorithm encrypts some of the parameters in the compressed video stream. Figure 5.14 shows the format information and texture information in MPEG1/2/4 compressed video stream. SECMPEG [25] is one of the well-known video stream encryption algorithms, which defines various levels for video stream encryption, including encrypting the whole stream, encrypting only the format information, and encrypting only I-frames, etc. This algorithm has the following properties:

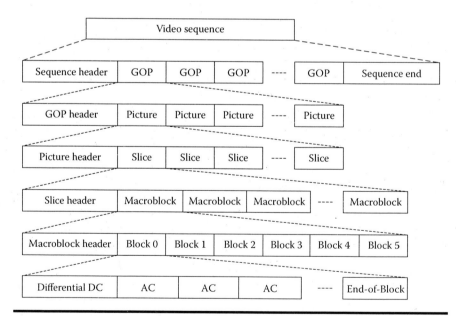

Figure 5.14 Format information and texture information in the compressed video.

- Encrypting the whole video stream obtains the highest security but the lowest encryption efficiency. Additionally, the format information is changed, and thus, the encrypted video cannot be displayed.

- Encrypting only format information is not secure. Since format information helps the decoder to recover the multimedia data, encrypting the format information will put the decoder out of work [26, 27]. However, it is not secure from a cryptographic viewpoint to encrypt only format information. This is because the format information is often in a certain grammar, which can be broken by statistical attacks [26]. Additionally, this algorithm changes the format information and thus makes the encrypted media data unable to be displayed or browsed by a normal browser.

- Encrypting only I-frames is not secure enough. In such video codecs as MPEG1/2/4, the frame is often classified into three types, I-frame, P-frame, and B-frame. I-frames are often encoded directly with DCT transformation, while P- and B-frames are often encoded by referencing to adjacent I- and P-frames. Thus, the I-frame is the referenced frame of P- and B-frames. Intuitively, encrypting only the I-frame will make the P- and B-frames unintelligible. However, experiments [26] show that this is not secure enough. The reason is that some macroblocks encoded with DCT transformation in P- and B-frames are left unencrypted. These are called intrablocks. For some videos with smart motion, the number of intrablocks is high enough to make the encrypted video intelligible.

- Encrypting only the intrablocks in I/P/B-frame is not secure enough. Generally, in a video sequence, more blocks are encoded by motion estimation and compensation than by DCT transformation, which are named interblocks. The intrablocks determine the texture information, and interblocks are more sensitive to motion. Thus, if the motion information is left unencrypted, the motion track in the video sequence may still remain intelligible.

5.5.2 Encryption during Compression

There exist many algorithms that encrypt the sensitive parameters during video compression. Taking MPEG2 and MPEG4 AVC/H.264, for example, the typical algorithms will be presented and analyzed below.

5.5.2.1 MPEG2 Video Encryption

In MPEG2 video encoding, for each block, the DCT block is produced by DCT transformation and quantization, and MVD is generated by motion estimation and compensation. Then, the DCT block is encoded by RLE (run-length encoding) and VLC (variable-length coding), and MVD is encoded by VLC, as shown in Figure 5.15. In this process, the DCT coefficients (DC and ACs) and MVD can be encrypted with various methods, such as sign encryption, DC encryption, DC and AC encryption, MVD encryption, etc.

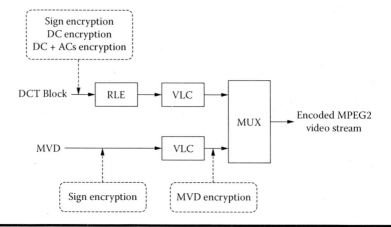

Figure 5.15   Video encryption in MPEG2 encoding.

The DCT block may be encrypted before RLE with sign encryption, DC encryption, DC and AC encryption. Here, sign encryption [28] means to extract the sign bits from the DCT block, encrypt the sign bits, and return them to the block. DC encryption [29] means to encrypt only the DC coefficient of the DCT block, while DC and AC encryption [30] means to encrypt both the DC coefficient and some AC coefficients in low frequency, as shown in Figure 5.16. According to the encoding process, sign encryption does not change the compression ratio, DC encryption changes the compression ratio slightly, and DC and AC encryption changes the compression ratio greatly. Among them, DC and AC encryption are more secure than the other two methods.

MVD can be encrypted before or after VLC. Before VLC, the MVD is encrypted with sign encryption. After VLC, the MVD is encrypted completely. For example, the encoded MVDs are modulated by the random sequence generated

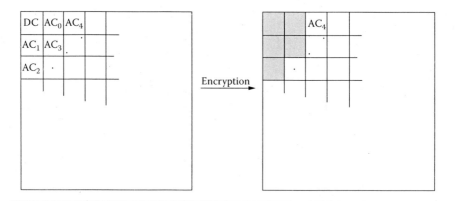

Figure 5.16   DC and AC encryption in MPEG2 encoding.

from a chaos system [31]. These methods do not change the compression ratio. Sign encryption is more efficient but less secure than MVD encryption. In most video codecs, a motion vector represents the motion information. Thus, encrypting the motion vector will make the motion track unintelligible. However, if there is little motion in a video, then few motion vectors are encrypted, and thus, the video is still intelligible. Therefore, encrypting only the motion vector is not secure enough.

Generally, both the MVD and DCT block are encrypted, in order to protect both the motion information and texture information. For example, in the algorithm proposed by Shi, Wang, and Bhargava [32], both the coefficient and motion vector signs are encrypted with DES or RSA [33]. Zeng and Lei [13] proposed an algorithm that permutes coefficients or encrypts the signs of coefficients and motion vectors in DCT or wavelet transformation.

The video is encrypted with various methods; the results are shown in Figure 5.17. Here, only the first P-frame of each encrypted video is listed. As can be seen, the videos encrypted by both DCT block encryption and MVD encryption are more chaotic than the ones encrypted by only DCT block encryption or MVD encryption. Generally, the methods combining block encryption and MVD encryption provide higher perceptual security.

Taking three kinds of methods for example, their encryption time ratios and compression ratio changes were tested and are shown in Table 5.1 and Table 5.2, respectively. The encryption methods tested include sign encryption, sign and DC encryption, and the encryption of sign, DC and the first 10 ACs. Here, the AES CTR cipher is used to encrypt the selected parameters, the methods are implemented in C program, and the computer is of 1.7 GHz CPU/256 M RAM. As can be seen, in the first two methods, the encryption time ratio between encryption and compression is always no more than 10%, while the encryption time ratio is sometimes bigger than 10% in the third method. Thus, the first two encryption methods can meet real-time requirements in practical applications. For the compression ratio change, there is a similar result. In the first method, the compression ratio remains unchanged, it changes slightly in the second method (generally smaller than 5%), and it changes greatly in the third method (bigger than 10%).

5.5.2.2   *MPEG4 AVC/H.264 Video Encryption*

In MPEG4 AVC/H.264 video encoding, for each block, the DCT block is produced by intra- and interprediction, DCT transformation and quantization, and MVD is generated by interprediction, motion estimation, and compensation. Then, the produced DCT block, IPM (Inter/intraprediction mode), and MVD are encoded by VLC (variable-length coding), as shown in Figure 5.18. In this process, the DCT coefficients (DC and ACs), IPM and MVD can be encrypted with various methods, such as sign encryption, DC encryption, VLC encryption, etc.

In the DCT block, the ACs are encrypted by sign encryption before VLC [34], the DC coefficients are encoded with VLC and encrypted completely, the

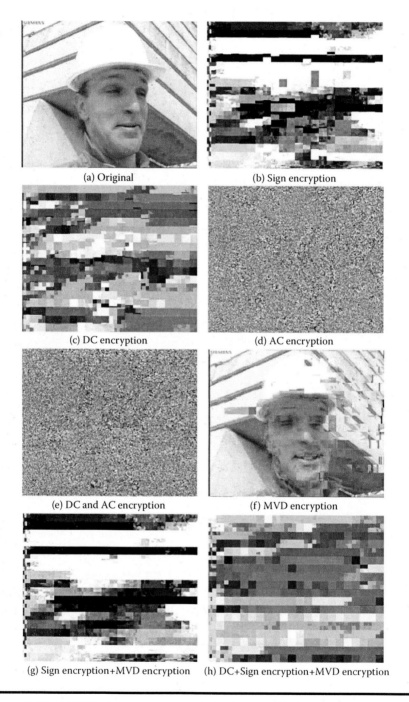

(a) Original   (b) Sign encryption

(c) DC encryption   (d) AC encryption

(e) DC and AC encryption (f) MVD encryption

(g) Sign encryption+MVD encryption (h) DC+Sign encryption+MVD encryption

Figure 5.17 MPEG2 videos encrypted by various methods.

Table 5.1 Test of Encryption Time Ratios

Video Sequence (I:P:B)	Encryption Time Ratio		
	Sign	*Sign + DC*	*Sign + DC + 10AC*
Bus (10:40:100)	0.8%	2.1%	11.1%
Bundy (8:35:81)	0.7%	2.2%	8.9%
Hulla (10:30:0)	1.1%	1.6%	9.2%
Car34 (6:6:22)	0.8%	2.0%	11.4%
Flower (8:38:96)	0.7%	2.0%	12.7%
Bike (10:40:98)	0.8%	1.8%	9.5%
Football (9:38:98)	0.7%	2.5%	13.8%
Salesman (10:40:100)	1.1%	2.3%	9.9%
Mobile (10:40:98)	0.9%	1.7%	10.7%

IPM is first encoded with VLC and then encrypted partially [34], and the MVD is encrypted with sign encryption before VLC. Among them, the sign encryption and complete encryption have been mentioned previously, while the partial encryption needs to be clarified. In AVC/H.264 codec, the IPM is encoded with Exp-Golomb codes [24]. This kind of codeword is composed of R zeros, one 1-bit and R bits of information (Y). Here, the intraprediction mode is $X = 2^R + Y - 1$, and $R = \lceil \log_2(X+1) \rceil$. The partial encryption process is shown in Figure 5.19, where the information part, Y, is encrypted with a cipher under the control of the key, and the other bits are left unchanged.

Generally, these encryption operations are combined to protect both the motion information and the texture information [34, 35]. Otherwise, using only one of them cannot confirm the security. For example, a scheme [36] is proposed to

Table 5.2 Test of Compression Ratio Changes in MPEG2 Video Encryption

Video Sequence (I:P:B)	Compression Ratio Change		
	Sign	*Sign + DC*	*Sign + DC + 10AC*
Bus (10:40:100)	0.0%	0.5%	19.7%
Bundy (8:35:81)	0.0%	0.3%	21.1%
Hulla (10:30:0)	0.0%	1.4%	24.3%
Car34 (6:6:22)	0.0%	1.2%	26.5%
Flower (8:38:96)	0.0%	0.8%	20.2%
Bike (10:40:98)	0.0%	1.1%	27.7%
Football (9:38:98)	0.0%	0.6%	31.5%
Salesman (10:40:100)	0.0%	0.7%	32.6%
Mobile (10:40:98)	0.0%	1.0%	24.2%

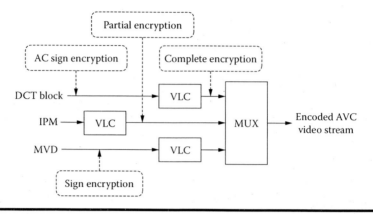

Figure 5.18 Video encryption in MPEG4 AVC/H.264 encoding.

permute only the intraprediction mode of each block, which degrades the video quality greatly. This algorithm is efficient in implementation, but is not secure against attacks. First, the motion information is left unencrypted, which makes the motion track still intelligible. Second, the video content can be recovered to some extent by replacement attacks [35]. Figure 5.20 shows the videos encrypted by MVD encryption, the method [35] combining all the encryption operations and the IPM permutation method [36]. Here, the first P-frame in each video sequence is listed. As can be seen, the combined method is more secure in perception. Additionally, the encryption time ratio is often smaller than 10%.

The computing complexity of the encryption methods depends on the data volumes to be encrypted and the cost of the adopted cipher. Among them, the data volumes to be encrypted include the intraprediction mode, the encoded DCs, the signs of ACs, and the signs of MVDs. Generally, the ratio of the encrypted data volumes is no bigger than 15%. The encryption time ratios of the methods were tested and are shown in Table 5.3. Here, the AES CTR cipher is used to encrypt the

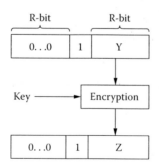

Figure 5.19 Partial encryption for IPM in MPEG4 AVC/H.264 encoding.

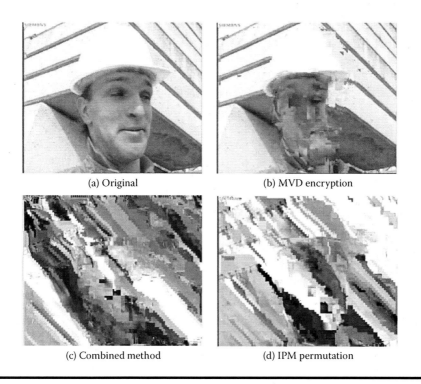

(a) Original

(b) MVD encryption

(c) Combined method

(d) IPM permutation

Figure 5.20   AVC/H.264 videos encrypted by different methods.

Table 5.3   Test of Compression Ratio Changes in AVC/ H.264 Video Encryption

		Encryption Time Ratio		
Video	*Size*	*MVD Encryption*	*IPM Permutation*	*Combined Method*
Foreman	QCIF	1.1%	0.4%	4.3%
Akiyo	QCIF	1.4%	0.7%	3.9%
Mother	QCIF	0.9%	0.6%	4.5%
Silent	QCIF	1.5%	0.8%	4.2%
News	QCIF	0.9%	0.5%	3.8%
Salesman	QCIF	0.8%	0.5%	3.4%
Mobile	CIF	1.9%	0.9%	5.1%
Football	CIF	1.2%	0.6%	3.7%
Akiyo	CIF	1.1%	0.5%	4.9%
Stephan	CIF	1.6%	0.9%	5.6%
Tempete	CIF	1.8%	1.0%	5.2%
Foreman	CIF	1.9%	1.0%	4.8%

selected parameters, the methods are implemented in C program, and the computer is of 1.7 GHz CPU/512 M RAM. As can be seen from the table, the encryption time ratio is no bigger than 10%. It means that the encryption/decryption operation does not affect the compression/decompression operation greatly. Additionally, these methods do not affect the compression ratio, which makes them practical for real-time applications.

5.6 Performance Comparison

Some partial encryption schemes have been presented and analyzed above. Their performances and application scenarios are listed in Table 5.4. For bit-plane encryption or object encryption, the security depends on the selected bit-planes or objects.

Table 5.4   Comparison of Various Partial Encryption Schemes

Encryption Algorithm	Security	Compression Ratio	Encryption Efficiency	Format Compliance	Application Suitability
Bit-plane encryption	SL0 or SL1	CL2	EL0	FL0	Multimedia storage
Object encryption	SL0 or SL1	CL2	EL0	FL0	Multimedia storage
Sign encryption in wavelet	SL2	CL0	EL0	FL2	Real-time interaction
Band permutation	SL2	CL2	EL0	FL2	Multimedia transmission
JPEG2000 stream encryption	SL0	CL0	EL0	FL2	Wireless/mobile communication
Mp3 encryption	SL0	CL0	EL0	FL1	Wireless/mobile communication
SECMPEG	SL0	CL0	EL1 or EL2	FL0 or FL1	Multimedia storage or transmission
DC encryption	SL3	CL1	EL0	FL1	None
DC+AC encryption	SL0	CL2	EL1	FL1	None
MVD encryption	SL3	CL0	EL0	FL1	None
DC + AC sign + MVD sign	SL0	CL1	EL0	FL1	Wireless/mobile communication
IPM permutation	SL3	CL0	EL0	FL1	None
Combined encryption for AVC/H.264	SL0	CL0	EL0	FL1	Wireless/mobile communication

For DC encryption, MVD encryption or IPM permutation, the encrypted video content is still intelligible to some extent, and thus, the algorithm is not secure in perception. They are not suitable for practical applications requiring security protection functionalities. In SECMPEG, various encryption modes are supported. For example, the complete encryption has the lowest efficiency and changes the file format completely. Encrypting only I-frames has higher efficiency and keeps some synchronization information unchanged.

5.7 Summary

In this chapter, some partial encryption algorithms for image, audio, and video data are presented and analyzed. They can be applied to the raw data, the coding parameters, or the compressed data stream. Their performances, including security, encryption efficiency, compression efficiency, and format compliance, are analyzed and compared. Additionally, typical application scenarios are proposed for different algorithms.

References

[1] ISO/MPEG-2. ISO 13818-2: Coding of moving pictures and associated audio, 1994.
[2] ISO/IECFCD15444-1: Information technology - JPEG2000 image coding system - Part 1: Core coding system, March 2000.
[3] H.264/MPEG4 Part 10 (ISO/IEC 14496-10): Advanced Video Coding (AVC), ITU-T H.264 standard.
[4] W. B. Pennebaker, and J. L. Mitchell. 1993. *JPEG Still Image Compression Standard*. NY: Van Nostrand Reinhold.
[5] ITU-T Recommendation H.263-Video Coding for Low Bit Rate Communication.
[6] J. M. Shapiro. 1993. Embedded image coding using zerotrees of wavelet coding. *IEEE Transactions on Signal Processing* 41(12): 3445–3463.
[7] A. Said. and W. A. Pearlman. 1996. A new fast and efficient image codec based on set partitioning in hierarchical trees. *IEEE Transactions on Circuits and Systems for Video Technology* 6(3): 243–250.
[8] H. Schwarz, D. Marpe, and T. Wiegand. 2007. Overview of the scalable video coding extension of the H.264/AVC Standard. *IEEE Transactions on Circuits and Systems for Video Technology* 17(9): 1103–1120.
[9] M. Podesser, H. P. Schmidt, and A. Uhl. 2002. Selective bitplane encryption for secure transmission of image data in mobile environments. In *Proceedings of the 5th IEEE Nordic Signal Processing Symposium (NORSIG 2002)* [CD-ROM], Tromso-Trondheim, Norway, October.
[10] K. S. Ntalianis, and S. D. Kollias. 2005. Chaotic video objects encryption based on mixed feedback, multiresolution decomposition and time-variant S-boxes. In *2005 IEEE International Conference on Image Processing (ICIP2005)*, September 11–14, 1110–1113.

[11] ZSoft PCX File Format Technical Reference Manual, http://www.qzx.com/pc-gpe/pcx.txt

[12] T. Uehara. 2001. Combined encryption and source coding. http://www.uow.edu.au/~tu01/CESC.html

[13] W. Zeng, and S. Lei. 2003. Efficient frequency domain selective scrambling of digital video. *IEEE Transactions on Multimedia* 5(1): 118–129.

[14] S. Lian, J. Sun, and Z. Wang. 2004. Perceptual cryptography on SPIHT compressed images or videos. *Proceedings of IEEE International Conference on Multimedia and Expo (I) (ICME 2004)*, Vol. 3, Taiwan, China, June, 2195–2198.

[15] S. Wee, and J. Apostolopoulos. 2003. *Secure Scalable Streaming and Secure Transcoding with JPEG-2000.* Technical Report, HPL-2003-117, June 19. Hp Laboratories, Palo Alto, CA.

[16] A. Pommer, and A. Uhl. 2003. Selective encryption of wavelet-packet encoded image data: Efficiency and security. *Communications and Multimedia Security* 194–204.

[17] R. Norcen, and A. Uhl. 2003. *Selective Encryption of the JPEG2000 Bitstream.* International Federation for Information Processing (IFIP). Lecture Notes in Computer Science, 2828, 194–204.

[18] S. Lian, J. Sun, D. Zhang, and Z. Wang. 2004. *A Selective Image Encryption Scheme Based on JPEG2000 Codec.* The 2004 Pacific-Rim Conference on Multimedia (PCM2004). Lecture Notes in Computer Science, 3332, 65–72.

[19] S. Sridharan, E. Dawson, and B. Goldburg. 1991. Fast Fourier transform based speech encryption system. *IEE Proceedings of Communications, Speech and Vision* 138(3): 215–223.

[20] A. Servetti, and J. C. Martin. 2002. Perception-based partial encryption of compressed speech. *IEEE Transactions on Speech and Audio Processing* 10(8): 637–643.

[21] A. Servetti, and J. C. Martin. 2002. Perception-based selective encryption of G. 729 speech. *Proceedings of IEEE ICASSP*, Vol. 1, Orlando, FL, 621–624.

[22] A. Servetti, C. Testa, J. Carlos, and D. Martin. 2003. Frequency-selective partial encryption of compressed audio. Paper presented at the International Conference on Audio, Speech and Signal Processing, Hong Kong, April.

[23] L. Gang, A. N. Akansu, M. Ramkumar, and X. Xie. 2001. Online music protection and MP3 compression. In *Proceedings International Symposium on Intelligent Multimedia, Video and Speech Processing*, May, 13–16.

[24] ITU-T Rec. H.264/ISO/IEC 11496-10. Advanced Video Coding. Final Committee Draft, Document JVT-E022, September 2002.

[25] L. Tang. 1996. Methods for encrypting and decrypting MPEG video data efficiently. In *Proceedings Fourth ACM International Multimedia Conference (ACM Multimedia'96)*, Boston, MA, November, 219–230.

[26] I. Agi, and L. Gong. 1996. An empirical study of MPEG video transmissions. In *Proceedings Internet Society Symposium on Network and Distributed System Security*, San Diego, CA, February, 137–144.

[27] C. H. Ho, and W. H. Hsu. 2005. System and method for image protection. TW227628B.

[28] C. Shi, and B. Bhargava. 1998. A fast MPEG video encryption algorithm. In *Proceedings 6th ACM International Multimedia Conference*. Bristol, UK, September, 81–88.

[29] A. N. Lemma, S. Katzenbeisser, M. U. Celik, and M. V. Veen. 2006. Secure watermark embedding through partial encryption. *Proceedings of International Workshop on Digital Watermarking (IWDW 2006)*. Lecture Notes in Computer Science, 4283, 433–445.

[30] A. Romeo, G. Romdotti, M. Mattavelli, and D. Mlynek. 1999. Cryptosystem architectures for very high throughput multimedia encryption: The RPK solution. In *Proceedings 6th IEEE International Conference on Electronics, Circuits and Systems (ICECS '99)*, Vol. 1, September 5–8, 261–264.

[31] J.-C. Yen, and J.-I. Guo. 1999. A new MPEG encryption system and its VLSI architecture. In *Proceedings IEEE Workshop on Signal Processing Systems*, Taipei, 430–437.

[32] C. Shi, S. Wang, and B. Bhargava. 1999. MPEG video encryption in real-time using secret key cryptography. In *Proceedings of Parallel and Distributed Processing Technologies and Applications*. Las Vegas, NV.

[33] R. A. Mollin. 2006. An Introduction to Cryptography. Boca Raton, FL: CRC Press.

[34] S. G. Lian, Z. X. Liu, Z. Ren, and H. L. Wang. 2006. Secure advanced video coding based on selective encryption algorithms. *IEEE Transactions on Consumer Electronics* 52(2): 621–629.

[35] S. Lian, Z. Liu, and Z. Ren. 2005. Selective video encryption based on advanced video coding. In *Proceedings 2005 Pacific-Rim Conference on Multimedia (PCM2005), Part II*. Lecture Notes in Computer Science, 3768, 281–290.

[36] J. Ahn, H. Shim, B. Jeon, and I. Choi. 2004. Digital video scrambling method using intra prediction mode. PCM2004. Lecture Notes in Computer Science, 3333, 386–393.

Chapter 6

Compression-Combined Encryption

6.1 Definition of Compression-Combined Encryption

Compression-combined encryption is the algorithm that combines the encryption operation and the compression operation and produces a secure data stream with smaller size. Compared with the traditional compression-encryption independent scheme shown in Figure 6.1(a), the combined encryption scheme realizes encryption and compression simultaneously, as shown in Figure 6.1(b).

The compression-combined encryption process can be described by Figure 6.2. Let A and B be the set of source data and compressed data, respectively, and they satisfy $A = \{P_0, P_1, \ldots, P_{N-1}\}$ and $B = \{C_0, C_1, \ldots, C_{N-1}\}$. Here, P_i is the ith plain-media ($i = 0, 1, \ldots, N-1$), C_i is the ith cipher-media, and P_i is compressed into C_i. That is, the following condition is satisfied.

$$C_i = \text{Com}(P_i).$$

Here, Com() is the compression function. C_i will be decompressed into P_i with the symmetric decompression function Decom().

$$P_i = \text{Decom}(C_i).$$

If the compression process can be controlled by the key K, then the compressed media will be changed, and thus, we get

$$C_j = \text{Com}(P_i, K).$$

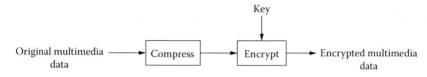

(a) Compression-encryption independent scheme

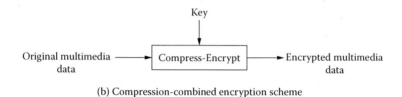

(b) Compression-combined encryption scheme

Figure 6.1 Architecture of compression-combined encryption.

Here, j may be an arbitrary number that depends on K. In decompression, if no K, the decompressed media is

$$P_j = \text{Decom}(C_i).$$

Thus, the decompressed media P_j is different from the original one P_i. And only the user with the K can recover the media correctly according to

$$P_i = \text{Decom}(C_i).$$

Thus, the encryption function is just the compression function controlled by the key.

Intuitively, to design a compression-combined encryption algorithm is just to add some encryption operations to the compression process. The difficulty is how to make the encryption operation not change the compression efficiency. Some encryption algorithms combined with the typical compression operations have been reported, such as Huffman coding, arithmetic coding, etc. They are introduced and analyzed in detail below.

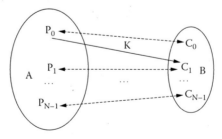

Figure 6.2 General description of compression-combined encryption.

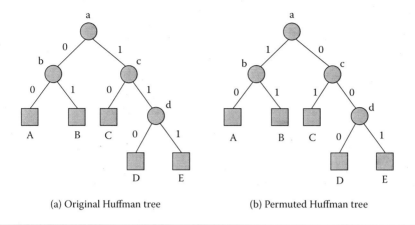

(a) Original Huffman tree   (b) Permuted Huffman tree

Figure 6.3   Random Huffman tree permutation.

6.2 Secure Huffman Coding Based on Random Huffman-Tree Permutation

Huffman coding [1] is one of the widely used entropy coding methods that adopt the statistical properties of the source data to obtain the ideal compression ratio. To encode source data composed of various symbols, the probabilities of the symbols occurring are computed, then the Huffman tree is constructed based on the computed probabilities; each symbol is assigned a codeword denoted by the path in the Huffman tree, and the source data is encoded symbol by symbol according to the table composed of the codewords. Here, the Huffman tree can be constructed adaptively, which is termed the adaptive Huffman tree [2]. According to entropy theory, the symbol with higher probability of occurring lies in the higher level of the Huffman tree. The codeword of the symbol is composed of the labels along the path from the root to the leaf node. Thus, the more often the symbol occurs, the shorter the symbol's codeword. Figure 6.3(a) shows the Huffman tree for symbols, A, B, C, D and E. The corresponding code table is shown in Table 6.1.

Table 6.1   Original Huffman Codebook

Symbol	Codeword
A	00
B	01
C	10
D	110
E	111

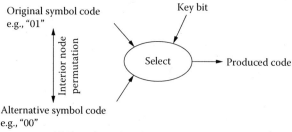

(a) The scheme based on interior node permutation

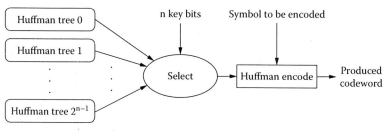

(b) The scheme based on Huffman tree permutation

Figure 6.4   Secure Huffman encoding.

Some means [3–5] have been proposed to shuffle or permute Huffman tree during Huffman encoding in order to realize secure encoding. Thus, only the decoder with the correct key can decode the data correctly. These means can be classified into two types, interior node permutation and Huffman tree permutation. In the former method, the interior node is permuted under the control of a key bit, as shown in Figure 6.4(a). For example, if A is to be encoded, and the key bit is "0", then the interior node b is not permuted, and A is encoded into "00". Otherwise, if A is to be encoded and the key bit is "1", then the interior node b is permuted, and A is encoded into "01". In this case, for each symbol, one key bit is required. Thus, for the source data composed of N symbols, the encryption space becomes 2^N. Here, only the encryption scheme combined with the fixed Huffman coding is considered, which can be extended to adaptive Huffman coding.

In another method, the original Huffman tree is permuted, which generates some new Huffman trees. The source data is encoded with reference to all the Huffman trees in a random manner. First, the Huffman tree permutation is realized by permuting the interior nodes. The number of permuted trees is determined by the number of interior nodes. If there are n interior nodes in a Huffman tree, the tree can be permuted into 2^n trees (including itself). Figure 6.3(b) shows an example permuting the interior nodes a and c. The corresponding code table is listed in Table 6.2. In this case, to encrypt each symbol, n key bits are required to select a Huffman tree or code table, as shown in Figure 6.4(b). Thus, for source data

Table 6.2   Permuted Huffman Codebook

Symbol	Codeword
A	10
B	11
C	01
D	000
E	001

composed of N symbols, the encryption space is 2^{nN}. As can be seen, this case is more secure against brute-force attack than the former one.

Secure Huffman coding encrypts the source data into an unintelligible form. Taking MPEG2 video, for example, the secure Huffman coding based on tree permutation is used to encode the discrete cosine transform (DCT) coefficients and motion vector differences (MVDs). As shown in Figure 6.5, the videos produced are too chaotic to be understood, which shows that the encryption method is secure in perception. However, considering that these encoding and encryption processes are constructed on codeword encryption, they are not secure against select-plaintext attack. Some means have been reported to improve the security, for example, using a random sequence to modulate the Huffman codes produced [5], using key hopping sequence to control the encryption process [6], and jointing random entropy coding and bitstream rotation [7]. However, the improvements on security are based on the sacrifices of some other performances, such as encryption efficiency or format compliance.

6.3 Secure Arithmetic Coding Based on Interval Permutation

Arithmetic coding [8–10] is also a well-known entropy coding method, which encodes the source data of any length as a real number between 0 and 1. The length of the source data determines the precision used in the coding. Longer source data

(a) Original video   (b) Encrypted video

Figure 6.5   Videos encrypted by secure Huffman encoding.

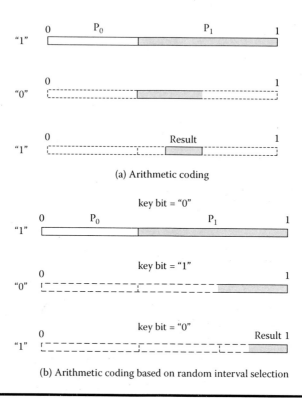

(a) Arithmetic coding

(b) Arithmetic coding based on random interval selection

Figure 6.6 Secure arithmetic coding.

is often encoded with more precision. Generally, arithmetic coding includes the following steps. First, the n symbols' probability distribution is computed. Second, the current interval is partitioned into n intervals according to the probability distribution (the initial interval is [0,1]). Third, the interval corresponding to the present symbol is selected as the current interval. Fourth, the next symbol is selected as the present symbol, and the steps from the second one to the fourth one are repeated until all the symbols are encoded. Finally, the current interval is changed into a binary form that is the coding result.

Taking source data composed of binary bits, for example, the arithmetic coding process is shown in Figure 6.6(a). Here, the source data are "1 0 1", the probability of "0" is p_0, and the probability of "1" is p_1. As can be seen, the result interval is $[p_0,1)$ after encoding the first bit "1", $[p_0,1-p^2_1)$ after encoding the second bit "0", and finally $[p_0 + p_1p^2_0, 1-p^2_1)$ after encoding the third bit "1". The decoding process decides the bits one by one with the help of the probability distribution.

Some means [3–5, 11] have been proposed to combine encryption with arithmetic coding. The first kind of algorithm changes the relationship of the intervals by randomly permuting them, and uses the permuted interval structure to encode the source data. The second one generates various interval structures by permuting

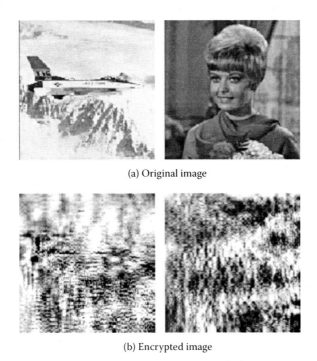

(a) Original image

(b) Encrypted image

Figure 6.7   JPEG2000 images encoded by secure arithmetic coding.

the intervals randomly, and randomly selects an interval structure to encode the source data. The third one encodes each symbol in the source data by randomly changing an interval's position. As can be seen, all of them need to permute or randomly change the interval. Compared with the normal coding in Figure 6.6(a), the third method is shown in Figure 6.6(b). Here, if the key bit is "1", the interval's position is changed. Otherwise, it is encoded with normal coding.

As can be seen, with the control of key bits, the result produced is quite different from the normal one. Thus, without the key, it is difficult to recover the source data. As an example, the secure arithmetic coding is used to replace the normal arithmetic coding in JPEG2000. The coded images are unintelligible, as shown in Figure 6.7. Note that the existing secure arithmetic coding is similar to the secure Huffman coding. Thus, it is also not secure against select-plaintext attacks [12–14].

6.4 Secure Coding Based on Index Encryption

Generally, there are existing schemes [15, 16] for codeword encryption. The typical one is table permutation, which randomly changes the original codeword table and uses the changed table to encode the source data. In decoding, the original table is

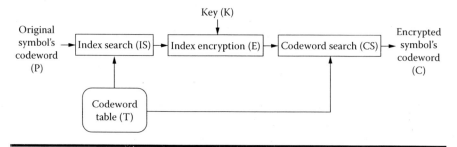

Figure 6.8 Secure fixed-length coding based on index encryption.

changed in the same manner and then used to decode the received data. For this scheme, the security depends on the size of the table, and it is not secure against known-plaintext attack. Another scheme is index encryption, which encrypts each symbol's codeword according to the codeword table. Index encryption based on fixed-length coding and variable-length coding are presented below.

6.4.1 Secure Fixed-Length Coding (FLC) Based on Index Encryption

Fixed-length coding (FLC) [17] transforms each symbol into a codeword of the same length. Index encryption is effective for FLC, which encrypts the codeword's index during or after symbol encoding. As shown in Figure 6.8, the secure encoding process is defined as

$$C = \text{CS}(\text{E}(\text{IS}(P,T),K),T).$$

Here, P, T, K and C are the original codeword, codeword table, key, and encrypted codeword, respectively, and, IS() and CS() are index search function and codeword search function, respectively. IS(P,T) means to get P's index in the table T, and CS(a,T) means to get a's codeword through indexing the table T.

The decryption process is symmetric to the encryption process. According to the encryption defined above, the decryption is defined as

$$P = \text{CS}(\text{D}(\text{IS}(C,T),K),T).$$

Here, the parameters and functions are the same as the ones used in encryption.

Because all the codewords in FLC have the same length, the index encryption operation does not change the data size. The key K may be fixed or variable with the symbols. If K is fixed for all the symbols, the encryption scheme is equivalent to table permutation, and the security against brute-force attack increases with

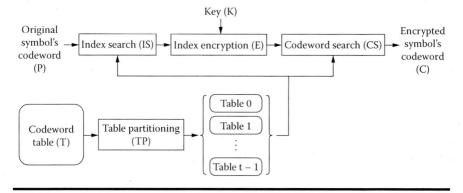

Figure 6.9 Secure variable-length coding based on index encryption.

the table size. However, it is not secure against known-plaintext or select-plaintext attack. Otherwise, if K is variable, the encryption scheme is equivalent to a stream cipher. Thus, to keep secure, K should vary with the symbols.

6.4.2 Secure Variable-Length Coding (VLC) Based on Index Encryption

Compared with FLC, variable-length coding (VLC) [18] transforms each symbol into a codeword of variable length. Index encryption can also be used for VLC, which encrypts the codeword's index during or after symbol encoding. However, considering that index encryption may change the size of the encoded data greatly, the codeword table (T) is first partitioned into subtables [15, 16], that is, Table 0, Table 1, … and Table $t-1$. Here, each subtable is composed of codewords with the same or nearly the same length. Then, each symbol is encoded and encrypted with the corresponding subtable. As shown in Figure 6.9, the secure encoding process is defined as

$$C = \mathrm{CS}(\mathrm{E}(\mathrm{IS}(P,T_i),K),T_i).$$

Here, P, K, C, IS() and CS() are the same as defined for secure FLC. T_i is the subtable that is selected according to the length of P. That is, T_i is the table containing the codewords with length equal to that of P.

The decryption process is symmetric to the encryption process. According to the encryption defined above, the decryption is defined as

$$P = \mathrm{CS}(\mathrm{D}(\mathrm{IS}(C,T_i),K),T_i).$$

Table 6.3   Exp-Golomb Code Table

Code_num	Signed Mapping	Codeword	
0	0	0	
1	1	010	T0
2	−1	011	
3	2	00100	
4	−2	00101	
5	3	00110	T1
6	−3	00111	
7	4	001000	
8	−4	0001001	
9	5	001010	T2
10	−5	0001011	
…	…	…	

Here, the parameters and functions are the same as the ones used in encryption. The subtable T_i is selected according to the length of C. That is, T_i is the table containing the codewords with length equal to that of C.

Because VLC is now used in media coding more widely than FLC, secure VLC is more practical for media encryption. Here, the secure VLC coding methods based on Exp-Golomb code and Huffman code are presented, as follows.

6.4.2.1 Secure Coding Based on Exp-Golomb Code

Exp-Golomb code [19] has been used to encode inter/intra-prediction mode (IPM) and MVD in advanced video coding (AVC) [20]. Table 6.3 shows the structures of the codewords. Here, the code_num or signed_num is encoded into the codeword composed of zeros and information. To realize secure encoding, the table is partitioned into subtables, as shown in Table 6.3. Thus, each code_num or signed_num is encoded and encrypted in the corresponding table, and the encryption key changes with time. For example, normally, code_num 4 or signed_num −2 is encoded as the codeword 00101. In secure coding, the codeword will be encoded in Table 1, and may be encoded as one of the four codewords, that is, 00100(3/2), 00101(4/−2), 00110(5/3) and 00111(6/−3).

The encrypted videos are often unintelligible, as shown in Figure 6.10. Because the codeword is encrypted into one with the same length, the compression ratio will not be changed. Additionally, the file format remains unchanged, and thus, the encrypted video content can still be displayed or edited, for example, displaying, GOP cutting, frame cutting, etc.

(a) Original video

(b) Encrypted video

Figure 6.10   Video encrypted by secure Exp-Golomb coding.

6.4.2.2 Secure Coding Based on Huffman Code

Huffman code is used in MPEG2 video coding [21], which encodes the DCT coefficients and MVDs. Taking coefficient encoding, for example, the coefficient's run-level is encoded with the Huffman table that is designed through various tests. To realize secure coding, the Huffman table should be partitioned into subtables. To partition the table strictly according to the codeword length is not suitable, because that will produce some tables containing few codewords. Thus, to improve security, the tables should be combined. The typical method is to construct five encryption tables according to codeword length, as shown in Table 6.4. T_0 is composed of the codewords with length varying from 2 to 4, T_1 is composed of the codewords with length varying from 5 to 7, T_2 is composed of the codewords with length varying from 8 to 10, T_3 is composed of the codewords with length of 11, and T_4 is composed of the codewords with length varying from 12 to 16. For example, the codeword 1111111110000010 is collected in T_4, as shown in Table 6.5.

The encrypted videos are often unintelligible, as shown in Figure 6.11. Because the codeword index can be encrypted by traditional cipher, the cryptographic

Table 6.4 Table Partitioning of Huffman Table

Symbol	Codeword
T_0	2~4
T_1	5~7
T_2	8~10
T_3	11
T_4	12~16

security can be confirmed. Additionally, the encryption scheme keeps the file format unchanged, and supports such direct operation as displaying, GOP cutting or frame cutting, etc. The disadvantage is that it changes the compression ratio slightly (generally, no more than 10%).

6.5 Random Coefficient Scanning

In some transform coding methods, such as EZW [22], SPIHT [23], MPEG2, MPEG4 AVC/H.264, etc., the transformed and quantized coefficients are scanned in a certain manner in order to adopt the adjacent relationship between coefficients. The random coefficient scanning method changes the coefficient scanning order during compression, which can degrade the quality of the media content greatly. Two typical methods suitable for popular compression codecs are presented in the following sections.

6.5.1 *Random Scanning in Wavelet Domain*

In a wavelet-based codec, such as EZW or SPIHT, the coefficients are encoded from the subband in the lowest frequency to the one in the highest frequency. At the same time, the quadtree-based relationship between the coefficient in the lower subband and the corresponding four coefficients in the higher subband is adopted

Table 6.5 Example of a Subtable

Index	Run, Level	Length (bits)	Codeword
0	0, 9	16	1111111110000010
1	0, A	16	1111111110000011
2	1, 6	16	1111111110000100
3	1, 8	16	1111111110000110
4	2, 4	12	111111110100
5	2, 7	16	1111111110001011
...	...	...	...

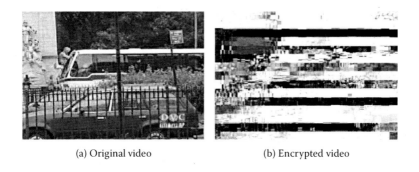

(a) Original video (b) Encrypted video

Figure 6.11 Video encrypted by secure Huffman coding.

to improve the compression efficiency. The normal scanning order is denoted by Figure 6.12(a). Here, the three-level wavelet transform is applied to an image.

The scanning order can be changed using one of three methods [24–26]. The first scans the coefficient in the whole image randomly, as shown in Figure 6.12(b). It is similar to the coefficient permutation method that permutes the coefficients in the whole image.

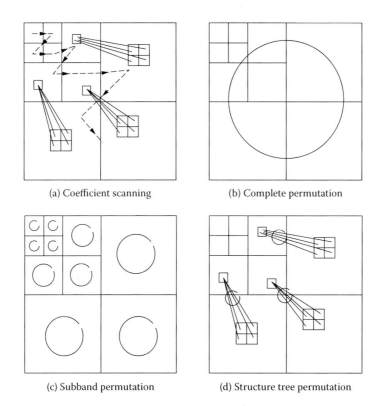

(a) Coefficient scanning (b) Complete permutation

(c) Subband permutation (d) Structure tree permutation

Figure 6.12 Random coefficient scanning in EZW/SPIHT encoding.

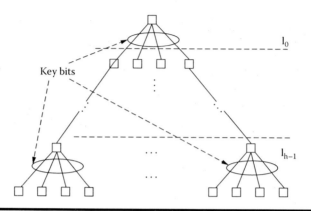

Figure 6.13   Quadtree permutation.

The second method scans the coefficient in each subband randomly, as shown in Figure 6.12(c). It is similar to the coefficient permutation method that permutes the coefficients in each subband. The third scans the coefficient in each quadtree randomly, as shown in Figure 6.12(d). It is similar to quadtree permutation, as shown in Figure 6.13.

In quadtree permutation [26], the relation between the node and its children nodes is changed. That is, the order of the children nodes is changed randomly under the control of key bits. Quadtree permutation is similar to Huffman tree permutation, with two differences. First, quadtree permutation permutes the positions of four nodes, whereas Huffman tree permutation permutes the positions of two nodes. Second, the quadtree in a wavelet codec is a complete quadtree, whereas the Huffman tree is not a complete tree. Thus, for the same height, quadtree permutation needs more key bits than Huffman tree permutation. Here, five bits are required to permute one interior node in a quadtree. For the h-height quadtree, there are 4^h interior nodes. Thus, 5×4^h key bits are required to permute the quadtree.

Figure 6.14 shows the images encrypted by various random scanning methods during SPIHT coding. As can be seen, the quadtree permutation is not secure in perception although it does not change the compression efficiency. Thus, it cannot be used to encrypt images independently [27].

6.5.2   Random Scanning in DCT Domain

In DCT-based codecs, such as MPEG2 or MPEG4 AVC/H.264, the DCT coefficients are scanned in a zigzag order that orders the coefficients from the lowest frequency to the highest frequency, as shown in Figure 6.15(a). To encrypt the coefficients, the scanning order can be changed by two methods [28, 29]. The first scans the coefficient in the whole DCT block randomly, as shown in Figure 6.15(b). It is similar to the coefficient permutation method that permutes the coefficients in the whole DCT block. The other scans the coefficient in each subband randomly, as shown in Figure 6.15(c). Here, the DCT block is partitioned into three segments

(a) Original image   (b) Complete permutation

(c) Subband permutation   (d) Quadtree permutation

Figure 6.14   Images encrypted by various random scanning methods.

according to the frequency. It is similar to the coefficient permutation method that permutes the coefficients in each segment.

Different random scanning methods have different permutation space, which determines the difficulty of brute-force attack. Taking the DCT block composed of N coefficients, for example, the permutation space for block permutation is

$$S_{Block} = 1 \cdot 2 \cdots N = N!.$$

In segment permutation, the block is partitioned into m segments, and each segment is composed of N_i ($i = 0, 1, \ldots, m - 1$) coefficients. Then, the permutation space satisfies

$$\begin{cases} N_0 + N_1 + \cdots + N_{m-1} = N (N \geq N_i \geq 2, i = 0, 1, \cdots, m-1) \\ S_{Segment} = (N_0!) \cdot (N_1!) \cdots (N_{m-1}!) \end{cases}.$$

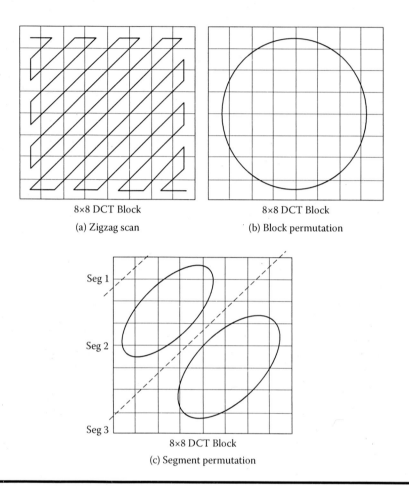

8×8 DCT Block

(a) Zigzag scan

8×8 DCT Block

(b) Block permutation

8×8 DCT Block

(c) Segment permutation

Figure 6.15   Random coefficient scanning in discrete cosine transform (DCT) block.

To compare sign encryption with random scanning, the encryption space of sign encryption is also analyzed here. As is known, in sign encryption, the sign bits 1 and 0 are changed randomly, which can be regarded as a permutation operation. For the block composed of N coefficients, the permutation space of sign encryption is

$$S_{Sign} = 2^N.$$

Then, it is easy to prove the following result.

$$S_{Sign} \leq S_{Segment} \leq S_{Block}.$$

Thus, among the three encryption methods, block permutation has the biggest permutation space, while sign encryption has the smallest permutation space. That

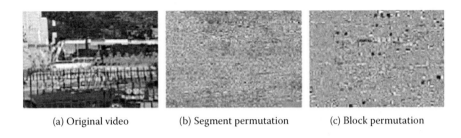

| (a) Original video | (b) Segment permutation | (c) Block permutation |

Figure 6.16   Videos encrypted by segment permutation and block permutation.

is, from a cryptographic perspective, block permutation has higher security than segment permutation and sign encryption.

Taking MPEG2 video coding, for example, each DCT block is encrypted with segment permutation or block permutation; the results are shown in Figure 6.16. Here, in segment permutation, the DCT block is partitioned into three segments, that is, [0, 4], [5, 19], and [20, 63]. As can be seen, the encrypted videos are all unintelligible. However, the compressed ratio is changed greatly, as shown in Table 6.6. Additionally, the permutation itself is not secure against known plaintext or select-plaintext attacks.

6.6 Performance Comparison

As mentioned above, various compression-combined encryption algorithms have been reported. These algorithms are compared and listed in Table 6.7 according to their performance in security, compression efficiency, encryption efficiency, format compliance, and application suitability. Because these algorithms combine with the corresponding compression operations, the file format remains unchanged.

Table 6.6   Compression Ratios Changed by Segment Permutation or Block Permutation

Video Sequence(I:P:B)	CCR[a] of Block Permutation (%)	CCR of Segment Permutation (%)
Bus(10:40:100)	35.7%	20.6%
Bundy(8:35:81)	38.2%	23.3%
Hulla(10:30:0)	32.4%	17.5%
Car34(6:6:22)	28.1%	15.7%
Flower(8:38:96)	34.3%	20.7%

[a] CCR, changed compression ratio.

Table 6.7 Performance Comparison of Various Compression-Combined Encryption Algorithms

Encryption Algorithm	Security	Compression Ratio	Encryption Efficiency	Format Compliance	Application Suitability
Secure Huffman coding based on Huffman tree permutation	SL2	CL0 or CL1	EL0	FL1	Wireless/mobile communication
Secure arithmetic coding based on interval permutation	SL2	CL0	EL0	FL1	Wireless/mobile communication
Secure FLC based on index encryption	SL0	CL0	EL1	FL1	Multimedia transmission
Secure VLC based on index encryption	SL0	CL0 or CL1	EL1	FL1	Multimedia transmission
Complete permutation	SL2	CL2	EL0	FL2	None
Subband permutation or segment permutation	SL2	CL2	EL0	FL2	None
Quadtree permutation	SL3	CL0	EL0	FL2	None

However, some of them change the compression efficiency, and most of them are not secure enough. For example, the random coefficient scanning in DCT domain often changes the compression ratio greatly. Although quadtree permutation does not change the compression ratio, it is not secure from either a perceptual or cryptographic viewpoint. Some other permutation methods may be secure in perception, but not secure against select-plaintext attack. Index encryption provides higher security because it is based on traditional cipher.

6.7 Summary

In this chapter, some compression-combined encryption algorithms are described and analyzed. From the analysis and comparisons, most of them are not secure enough. Especially, some algorithms are constructed based only on permutation operations. They should be improved before being used. Otherwise, they should be used together with some other strong algorithms.

References

[1] R. G. Gallager. 1978. Variations on a theme by Huffman. *IEEE Transactions on Information Theory* 24(6): 668–674.

[2] G. V. Cormack, and R. N. Horspool. 1984. Algorithms for adaptive Huffman codes. *Information Processing Letter* 18(3): 159–165.

[3] C. Wu, and C.-C. J. Kuo. 2001. Efficient multimedia encryption via entropy codec design. Paper presented at SPIE International Symposium on Electronic Imaging, San Jose, CA, January.

[4] C. Wu, and C.-C. J. Kuo. 2000. Fast encryption methods for audiovisual data confidentiality. Paper presented at SPIE Photonics East - Symposium on Voice, Video, and Data Communications, Boston, MA, November.

[5] C.-E. Wang. n.d. Cryptography in Data Compression. http://citeseer.ist.psu.edu/642122.html.

[6] D. Xie, and C.-C. J. Kuo. 2004. Enhanced multiple Huffman table (MHT) encryption scheme using key hopping. In *Proceedings IEEE International Symposium on Circuits and Systems (ISCAS 2004)*, Vancouver, Canada, May 23-26, 568–571.

[7] D. Xie, and C.-C. J. Kuo. 2007. Multimedia encryption with joint randomized entropy coding and rotation in partitioned bitstream. *EURASIP Journal on Information Security* Vol. 2007, Article ID 35262.

[8] C. B. Jones. 1984. An efficient coding system for long source sequences. *IEEE Transactions on Information Theory* 27: 280–291.

[9] A. Moffat, R. M. Neal, and I. H. Witten. 1995. Arithmetic coding revisited. *ACM Transactions on Information Systems* 16: 256–294.

[10] I. H. Witten, R. M. Neal, and R. J. Cleary. 1987. Arithmetic coding for data compression. *Communications of the ACM* 30: 520–540.

[11] H. A. Bergen, and J. M. Hogan. 1992. Data security in a fixed-model arithmetic coding compression algorithm. *Computers and Security* 11(5): 445–461.

[12] H. A. Bergen, and J. M. Hogan. 1993. A chosen plaintext attack on an adaptive arithmetic coding compression algorithm. *Computers and Security* 12: 157–167.

[13] M. Grangetto, E. Magli, and G. Olmo. 2006. Multimedia selective encryption by means of randomized arithmetic coding. *IEEE Transaction on Multimedia,* 8(5): 905–917.

[14] J. G. Clearly, S. A. Irvine, and I. Rinsma-Melchert. 1995. On the insecurity of arithmetic coding. *Computers and Security* 14: 167–180.

[15] J. Wen, M. Sevra, M. Luttrell, and W. Jin. 2001. A format-compliant configurable encryption framework for access control of multimedia. In *Proceedings IEEE Workshop on Multimedia Signal Processing,* Cannes, France, October, 435–440.

[16] S. Lian, Z. Liu, Z. Ren, and H. Wang. 2006. Secure distribution scheme for compressed data streams. In *Proceedings IEEE International Conference on Image Processing (ICIP 2006),* 1953–1956.

[17] The Unicode Standard 4.1, formally defined as UTF-32. http://www.unicode.org/versions/Unicode4.0.0/ch03.pdf

[18] Variable Length Coding (VLC), http://en.wikipedia.org/wiki/Variable-length_code

[19] Exp-Golomb code, http://en.wikipedia.org/wiki/Exponential-Golomb_coding

[20] H.264/MPEG4 Part 10 (ISO/IEC 14496-10): Advanced Video Coding (AVC), ITU-T H.264 standard.

[21] ISO/MPEG-2. ISO 13818-2: Coding of moving pictures and associated audio, 1994.

[22] J. M. Shapiro. 1993. Embedded image coding using zerotrees of wavelet coding. *IEEE Transactions on Signal Processing* 41(12): .

[23] A. Said and W. A. Pearlman. 1996. A new fast and efficient image codec based on set partitioning in hierarchical trees. *IEEE Transactions on Circuits and Systems for Video Technology* 6(3): 243–258.

[24] Uehara, T. Combined encryption and source coding. http://www.uow.edu.au/~tu01/CESC.html, 2001.

[25] W. Zeng, and S. Lei. 2003. Efficient frequency domain selective scrambling of digital video. *IEEE Transactions on Multimedia* 5(1): 118–129.

[26] S. Lian, J. Sun, and Z. Wang. 2004. Perceptual cryptography on SPIHT compressed images or videos. In *Proceedings IEEE International Conference on Multimedia and Expro (I) (ICME 2004),* Vol. 3, Taiwan, China, June, 2195–2198.

[27] S. Lian, J. Sun, and Z. Wang. 2004. A secure 3D-SPIHT codec. In *Proceedings European Signal Processing Conference,* September 6-10, Vienna, Austria, 813–816.

[28] L. Tang. 1996. Methods for encrypting and decrypting MPEG video data efficiently. In *Proceedings Fourth ACM International Multimedia Conference (ACM Multimedia'96),* Boston, MA, November, 219–230.

[29] A. S. Tosun, and W.-C. Feng. 2000. Efficient multi-layer coding and encryption of MPEG video streams. In *Proceedings IEEE International Conference on Multimedia and Expo,* Vol. 1, July 30–August 2, 119–122.

Chapter 7

Perceptual Encryption

7.1 Definition of Perceptual Encryption

Perceptual encryption is the encryption algorithm that degrades the quality of media content according to security or quality requirements. The typical application is secure media content preview. That is, the media content is degraded by perceptual encryption. Thus, although the media content is of low quality, the customer can still understand it. If he is interested in it, he will pay for a high-quality copy.

Generally, perceptual encryption is realized by encrypting some sensitive parameters. Thus, it belongs to the category partial encryption. There are two core issues in perceptual encryption. The first is how to partition media data into sensitive parameters. Here, the significance denotes the sensitivity to human perception. The second issue is how to select the sensitive parameters according to security or quality requirements. There are some means [1–7] to accomplish this. Suppose there are N parameters in media data P, and the parameters are ordered from the least significant to the most significant that is, $p_0, p_1, \ldots, p_{N-1}$. Let the quality factor Q range from 0 to 100, with the higher value for Q denoting better quality. Then the number of parameters that will be encrypted is determined by

$$n = \left\lfloor \frac{(100-Q)N}{100} \right\rfloor,$$

where $\lfloor x \rfloor$ denotes the highest number no larger than x. Additionally, the n parameters are selected from the order 0 to n, as shown in Figure 7.1. Thus, the second issue can be solved.

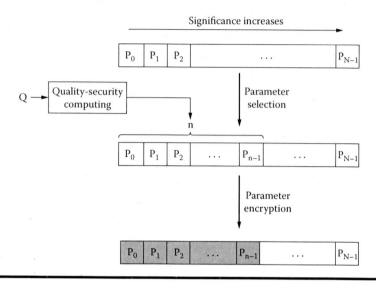

Figure 7.1   General architecture of perceptual encryption.

For the first issue, of partitioning media data into sensitive parameters, there are still no general solutions. According to the properties of existing media data and their compression methods, the parameter sensitivity can be classified according to the metrics list in Table 7.1. The first method uses the color-gray property as the metric. For colorful image or video data, the color information provides more information than gray information from the perspective of commercial value. The second method classifies the parameters according to their importance to decoding. For example, in an mp3 data stream, the bit allocation information and scale factor are more important than some coefficients because they determine the decoding

Table 7.1   Parameters Partitioning According to Sensitivity or Significance

Method	Metric of Partitioning	Suitability
1	Color-gray	Colorful image or video data
2	Parameters	Compressed data stream (mp3)
3	Bit-planes	Raw data, bit-plane-based codecs (JBIG, JPEG2000, MPEG4 FGS)
4	Frequency band	Transform coding (DCT, wavelet, MDCT, etc.)
5	Resolution	Progressive or scalable codecs (JPEG2000, SVC)
6	Progressiveness or scalability	Scalable data stream (including methods 3, 4, 5)

process. The third method uses the bit-planes as the metric. As is known, for an image, the significant bit-planes are preferred to be encoded first over the less significant ones, which also is used in such codecs as JBIG [8], JPEG2000 [9], MPEG4 FGS [10], etc. The third method partitions media data according to the frequency band. Generally, in such transformations as discrete cosine transform (DCT) [11], wavelet [12] or modified discrete cosine transform (MDCT) [13], most of the energy is concentrated on a low frequency band that determines the media content's approximate information, while the higher frequency band contains detailed information. The fifth method considers the resolution in the spatial or temporal domain. For example, taking an image after it is downsampled, the quality is degraded. Thus, some codecs make use of the resolution property to control the media quality. The sixth method uses the progressiveness or scalability in existing scalable data streams. The data stream may be encoded by such codecs as JPEG2000, MPEG4 FGS, SPIHT [14], SVC [15], etc. The scalability includes the resolution and quality, and the quality depends on the bit-planes or frequency band. Therefore, the sixth method can be regarded as a combination of the third, fourth and fifth methods.

After selecting the first n parameters, the original media content P composed of $p_0, p_1, \ldots, p_{N-1}$ is encrypted according to

$$c_i = \mathrm{E}(p_i, K), \; i = 0, 1, \ldots, n-1,$$

where c_i, K, $\mathrm{E}()$ are the ith ($i = 0, 1, \ldots, n-1$) cipher-parameter, key and encryption algorithm, respectively. Typically, different parts can be encrypted with different keys. Thus, the encrypted media content C is composed of $c_0, c_1, \ldots, c_{n-1}$ and $p_n, p_{n+1}, \ldots, p_{N-1}$. Similarly, in decryption, only the first n parts are decrypted according to

$$p_i = \mathrm{D}(c_i, K), \; i = 0, 1, \ldots, n-1,$$

where K and $\mathrm{D}()$ are the key and decryption algorithm, respectively. Typically, different parts can be decrypted with different keys. Thus, the media content P composed of $p_0, p_1, \ldots,$ and p_{N-1} can be recovered from C.

Two typical methods, the bit-plane-based method and the frequency band-based method are presented and analyzed below.

7.2 Perceptual Encryption Based on Bit-Plane Selection

7.2.1 Bit-Plane Encryption in the Spatial Domain

For raw data, the bit-planes are partitioned according to the image pixels. The most significant bit-plane is composed of all the image pixels' most significant bits, the least significant bit-plane is composed of all the image pixels' least significant bits, and the other bit-planes between the most significant one and the least significant one are composed of the pixels' corresponding bits. Thus, the size of the bit-plane depends on the size of the image. Taking an 8-bit image, for example, the bit-planes are partitioned according to the method shown in Figure 7.2.

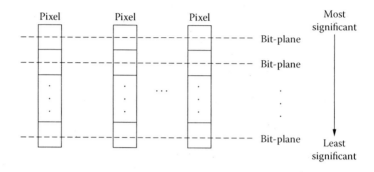

Figure 7.2   Bit-plane partitioning in an 8-bit image.

Different bit-planes have different significance for the quality of media content. According to the perceptual encryption defined in Section 7.1, the 8 bit-planes act as the parameters, that is, $N = 8$. The quality of the encrypted images corresponding to different n ($n = 1, 2, \ldots, 8$) are tested. Figure 7.3 shows the encrypted images and Figure 7.4 shows the relation between n and the peak signal-to-noise ratios (PSNRs) of the encrypted images. As can be seen, with increase in n, the quality of the images decreases. Generally, the encrypted image is unintelligible when n is no smaller than 7.

7.2.2 Bit-Plane Encryption in the DCT Domain

Some media compression methods encode the DCT coefficients bit-plane by bit-plane in order to obtain a trade-off between compression ratio and quality, such as

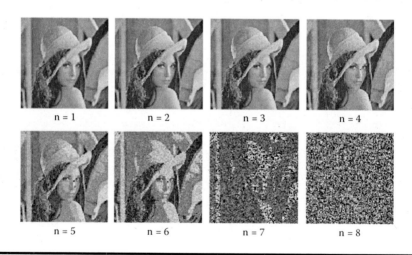

Figure 7.3   Images encrypted by bit-plane encryption.

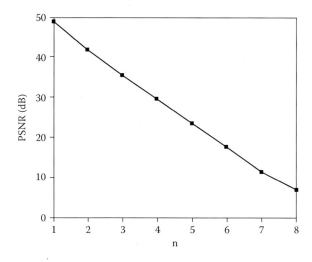

Figure 7.4   Relation between the quality of the encrypted images and *n*.

JBIG and MEPG4 FGS. In these codecs, the bit-plane is partitioned according to DCT coefficients. Considering that an 8×8 DCT transformation is often used in these codecs, the size of the bit-plane is 8×8, which is different from the spatial domain partitioning.

As above, different bit-planes have different significance for the quality of the media content. Taking $N = 9$ for example, the quality of the encrypted images corresponding to different n ($n = 1, 2, ..., 8$) is tested. Figure 7.5 shows the encrypted images, and Figure 7.6 shows the relation between n and the PSNRs of the encrypted

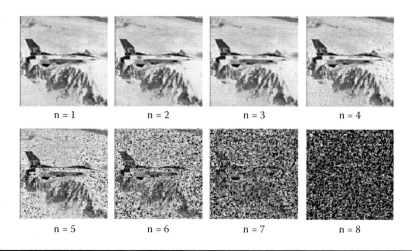

Figure 7.5   Images encrypted by bit-plane encryption in a DCT block.

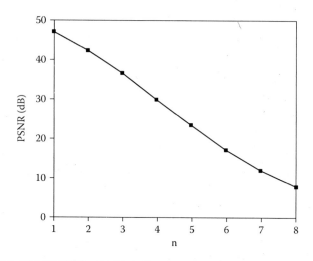

Figure 7.6 Relation between the quality of the encrypted images and *n*.

images. As can be seen, with increase in *n*, the quality of the images decreases. Generally, the encrypted image is unintelligible when *n* is no smaller than 7.

7.2.3 Bit-Plane Encryption in the Wavelet Domain

Some media compression methods encode the wavelet coefficients bit-plane by bit-plane in order to obtain a trade-off between compression ratio and quality, such as SPIHT and JPEG2000. In these codecs, the bit-plane is partitioned according to wavelet coefficients. Considering that the whole image is transformed by wavelet transformation, the bit-plane's size is the same as the image size.

Taking $N = 11$ for example, the quality of the encrypted images corresponding to different *n* ($n = 1, 2, ..., 8$) is tested. Figure 7.7 shows the encrypted images, and Figure 7.8 shows the relation between *n* and the PSNRs of the encrypted images. As can be seen, with increase in *n*, the quality of the images decreases. Generally, the encrypted image is unintelligible when *n* is no smaller than 8.

7.3 Perceptual Encryption Based on Frequency Band Selection

7.3.1 Frequency Band Encryption in a DCT Block

In a DCT-based codec, media data are partitioned into blocks (typically, 8×8 or 4×4), and each block is transformed by DCT, quantized and encoded with entropy coding. Generally, the DCT block is scanned in zigzag order, which generates the

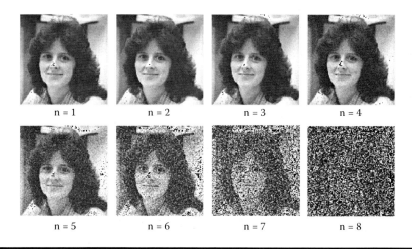

Figure 7.7   Images encrypted by bit-plane encryption.

coefficient sequence ordered from the lowest frequency to the highest frequency (from top-left to bottom-right). In this coefficient sequence, the first coefficient denotes the DCT block's energy, and the other coefficients denote detailed information on the image block. In order to realize perceptual encryption, the zigzag scan is applied inversely, that is, the DCT block is scanned from the bottom-right to the top-left, as shown in Figure 7.9. Thus, the last coefficient in the coefficient sequence denotes the block's energy.

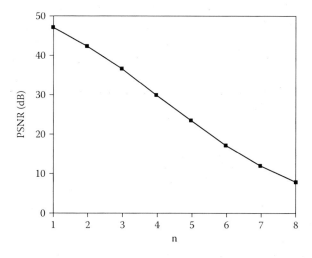

Figure 7.8   Relation between the quality of the encrypted images and *n*.

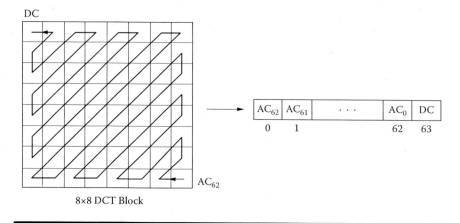

Figure 7.9   Coefficient sequence generation in a DCT block.

In perceptual encryption, the 64 coefficients can be selected from the first one to the last one according to the quality factor Q. Thus, set $N = 64$ for each DCT block. The quality of the encrypted images corresponding to different n ($n = 1, 2, \dots, 64$) is tested. Figure 7.10 shows the encrypted images, and Figure 7.11 shows the relation between n and the PSNRs of the encrypted images. As can be seen, with increase in n, the quality of the images decreases. Generally, the encrypted image is unintelligible when n is no smaller than 36.

7.3.2 Frequency Band Encryption in the Wavelet Domain

In a wavelet-based codec, media data is transformed by wavelet function, quantized and encoded with entropy coding. Generally, the transformed media data is partitioned into

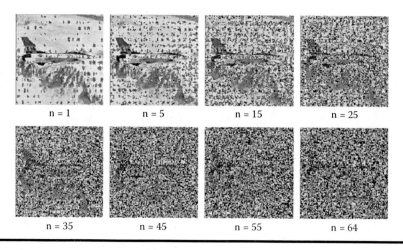

Figure 7.10   Images encrypted by DCT encryption.

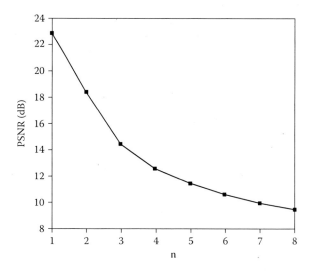

Figure 7.11   Relation between the quality of the encrypted images and *n*.

several frequency bands. The frequency bands are scanned in zigzag order, which generates a band sequence ordered from the highest frequency band to the lowest frequency band. Generally, the lower frequency band denotes approximate information, while the higher frequency band denotes detailed information. Figure 7.12 shows the frequency band sequence generated from the three-level wavelet transformation.

In perceptual encryption, the frequency band can be selected from the first to the last according to the quality factor Q. Taking five-level wavelet transformation, for example, the number of frequency bands is $N = 16$. Figure 7.13 shows the

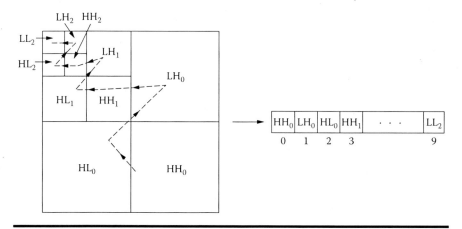

Figure 7.12   Band sequence generation in a wavelet domain.

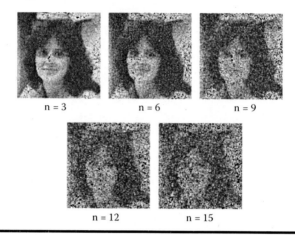

n = 3   n = 6   n = 9

n = 12   n = 15

Figure 7.13   Images encrypted by wavelet encryption.

encrypted images corresponding to different *n*, and Figure 7.14 shows the relation between *n* and the PSNRs of the encrypted images. As can be seen, with increase in *n*, the quality of the images decreases. Generally, the encrypted image is unintelligible when *n* is no smaller than 13.

7.4   Performance Comparison

According to the above analysis, perceptual encryption is constructed based on partial encryption, which reduces perceptual security by reducing the encrypted data volumes. To get higher perceptual security, more sensitive data parameters

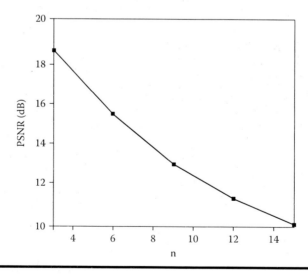

Figure 7.14   Relation between the quality of the encrypted images and *n*.

should be encrypted. To encrypt the selected data parameters, the encryption algorithms described in the section on partial encryption can be adopted. For example, the bit-planes in raw data can be directly encrypted by traditional ciphers, the bit-planes in the DCT or wavelet domain can be encrypted before or after they are compressed, and the frequency bands in DCT or wavelet can be encrypted during or after compression. If the bit-planes or frequency bands are encrypted before or during compression, the compression efficiency will be changed. The encryption efficiency depends on the number of the encrypted bit-planes or frequency bands. Generally, these algorithms can be used in real-time interaction, such as secure media content preview.

7.5 Summary

In this chapter, perceptual encryption is defined and introduced, and some typical perceptual encryption algorithms are analyzed. They are bit-plane encryption for raw data, bit-plane encryption in the DCT block or wavelet domain, and frequency band encryption in DCT block or wavelet domain. These algorithms are constructed on partial encryption. Thus, their cryptographic security is determined by the corresponding partial encryption algorithms. But the perceptual security varies with the number of encrypted parameters. Generally, the more the parameters are encrypted, the higher the perceptual security is. Additionally, their encryption efficiency also depends on the number of encrypted parameters. The typical application is secure media content preview during real-time interaction.

References

[1] A. Torrubia, and F. Mora. 2003. Perceptual cryptography of JPEG compressed images on the JFIF bit-stream domain. In *Proceedings IEEE International Symposium on Consumer Electronics, ISCE*, June 17–19, 58–59.

[2] A. Torrubia, and F. Mora. 2002. Perceptual cryptography on MPEG Layer III bitstreams. In *Proceedings IEEE International Conference on Consumer Electronics*, June 18–20, 324–325.

[3] A. Torrubia, and F. Mora. 2002. Perceptual cryptography on MPEG Layer III bitstreams. *IEEE Transactions on Consumer Electronics* 48(4): 1046–1050.

[4] S. Lian, J. Sun, and Z. Wang. 2004. Perceptual cryptography on SPIHT compressed images or videos. In *Proceedings IEEE International Conference on Multimedia and Expro (I) (ICME2004)*, Vol. 3, Taiwan, 2195–2198.

[5] S. Lian, Z. Liu, Z. Ren, and Z. Wang. 2007. Multimedia data encryption in block based codecs. *International Journal of Computers and Applications* 29(1): 18–24.

[6] S. Lian, X. Wang, J. Sun, and Z. Wang. 2004. Perceptual cryptography on wavelet-transform encoded videos. In *Proceedings 2004 International Symposium on Intelligent Multimedia, Video and Speech Processing (ISIMP'2004)*, October 20–22, Hong Kong 57–60.

[7] S. Li, G. Chen, A. Cheung, B. Bhargava, and K.-T. Lo. 2007. On the design of perceptual MPEG-Video encryption algorithms. *IEEE Transactions on Circuits and Systems for Video Technology* 17(2): 214–223.

[8] JBIG lossless image compression standard, ISO/IEC standard 11544 and ITU-T recommendation T.82.

[9] ISO/IECFCD15444-1: Information technology - JPEG2000 image coding system - Part 1: Core coding system, March 2000.

[10] W. Li. 2001. Overview of fine GRANULARITY SCALABILITY in MPEG-4 video standard. *IEEE Transactions on Circuits and Systems for Video Technology* 11(3): 301–317.

[11] S. A. Khayam. 2003. The Discrete Cosine Transform (DCT): Theory and Application. Unpublished paper. Michigan State University, Department of Electrical and Computer Engineering, http://www.egr.msu.edu/waves/people/Ali_files/DCT_TR802.pdf

[12] P. S. Addison. 2002. *The Illustrated Wavelet Transform Handbook*. London: Institute of Physics.

[13] V. Nikolajevic, and G. Fettweis. 2003. Computation of forward and inverse MDCT using Clenshaw's recurrence formula. *IEEE Transactions on Signal Processing* 51(5): 1439–1444.

[14] A. Said and W. A. Pearlman. 1996. A new fast and efficient image codec based on set partitioning in hierarchical trees. *IEEE Transactions on Circuits and Systems for Video Technology* 6(3): 243–250.

[15] H. Schwarz, D. Marpe, and T. Wiegand. 2007. Overview of the scalable video coding extension of the H.264/AVC Standard. *IEEE Transactions on Circuits and Systems for Video Technology* 17(9): 1103–1120.

Chapter 8

Scalable Encryption

8.1 Definition of Scalable Encryption

Scalable encryption is the encryption algorithm that protects scalable media streams and supports the direct operation of bit rate conversion. A scalable media stream [1, 2] is composed of media content with different scalability, including resolution in the spatial or temporal domain and the quality. Generally, media content can be changed by directly cutting some parts from the stream, and no recompression is required. Scalable encryption [3–8] is used to protect the scalable media stream while maintaining the scalability of the stream.

The general application scenarios are shown in Figure 8.1. In the first case without encryption, the media content is compressed into the media stream with scalable media encoding, then the media stream's size is changed by bit rate conversion, and finally, the converted media stream is decompressed with scalable media coding. In the second case with encryption, the encryption operation is inserted between the compression operation and bit rate conversion operation. Because the encrypted media stream can still be recovered after bit rate conversion, the encryption operation is called scalable encryption.

According to the property of scalable media stream, scalable encryption should satisfy two properties.

- The media content is protected in a secure manner. The encryption scheme is secure against cryptographic attacks, and the encrypted media content is too chaotic to be understood.
- The encrypted media stream can be directly operated without decryption and decoding. For example, the encrypted media stream can be directly cut in order to adapt the bit rate, and the remaining media stream can still be decrypted.

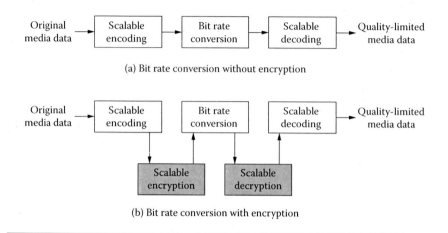

(a) Bit rate conversion without encryption

(b) Bit rate conversion with encryption

Figure 8.1   General application scenario of scalable encryption.

To design suitable scalable encryption algorithms, the properties of scalable coding should be considered. Thus, various scalable coding methods are investigated, and the corresponding scalable encryption algorithms are presented.

8.2   Scalable Media Coding

There exist some media encoding methods that have some scalability. Generally, the scalabilities of quality, image sizes, or frame rates, refer to SNR (signal-to-noise ratio), spatial, or temporal scalability, respectively. According to the scalable property, they can be classified into four types, as shown in Table 8.1. They are layered coding, layered and progressive coding, progressive coding and scalable coding.

Table 8.1   Classification of Scalable Coding Methods

Scalable Coding Method	Scalability	Corresponding Codecs
Layered coding	Nonscalable layers	MPEG2, H.263
Layered and progressive coding	Nonscalable base layer	MPEG4 FGS
	Scalable enhancement layer	
Progressive coding	Scalable bitstream	EZW, SPIHT, JBIG
Scalable coding	Scalable and reordered data stream	JPEG2000, SVC

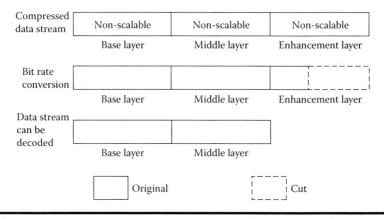

Figure 8.2 The data stream in layered coding.

8.2.1 Layered Coding

The first method, layered coding, compresses video data into multiple layers. The compressed layers are classified into three types, a base layer, a middle layer, and an enhancement layer. Among them, the base layer can be independently decoded and provide coarse visual quality, while the middle layer and the enhancement layer can only be decoded with reference to the base layer and can provide better visual quality. If all the layers are decoded, the resulting video will be of the highest quality. Otherwise, decoding the base layer or multiple layers will provide video with degraded quality, or a smaller image size or a lower frame rate. Additionally, the base layer, middle layer, and enhancement layer are all nonscalable, as shown in Figure 8.2. This means that if some parts of the base layer, middle layer or enhancement layer are cut off, the remaining portion cannot be decoded correctly. Thus, each layer is either entirely decoded or not decoded. The typical codecs belonging to this method are MPEG2 [9] and H.263 [10].

8.2.2 Layered and Progressive Coding

The second method, layered and progressive coding, is similar to layered coding. The main difference is that it produces a nonscalable base layer and a scalable enhancement layer, as shown in Figure 8.3. The base layer will be entirely decoded or not decoded, while the enhancement layer can be truncated. The typical codec using this encoding method is MPEG4 FGS (Fine Granularity Scalability) [11]. In MPEG4 FGS, the base layer is encoded with the traditional nonscalable coder. The enhancement layer denotes the difference between the original frame and the reconstructed frame, which is encoded by bit-plane encoding of discrete cosine transform (DCT) coefficients. In bit rate conversion, the bitstream of the enhancement layer can be truncated into any number of bits directly. The decoder can

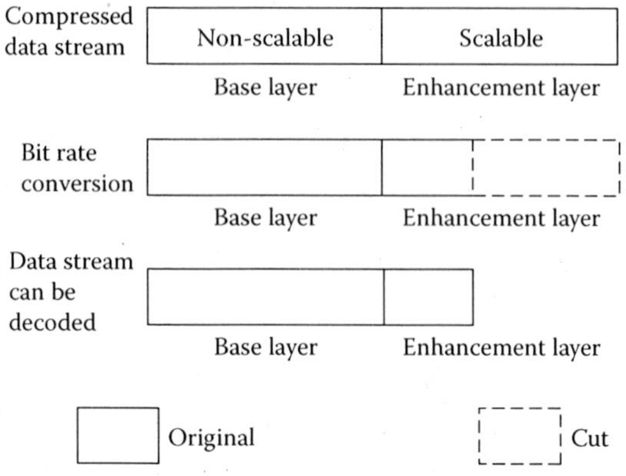

Figure 8.3   The data stream in layered and progressive coding.

reconstruct the video from the base layer and the truncated enhancement layer. The quality of the decoded video is proportional to the number of bits decoded in the enhancement layer. Additionally, FGS is combined with temporal scalability to support both quality scalability and temporal scalability.

8.2.3 Progressive Coding

The third method, progressive coding, compresses media data into a progressive bitstream. Generally, the bitstream has only one of the scalabilities, such as SNR scalability, and cannot be reordered according to other scalabilities. But the bitstream can be truncated directly, as shown in Figure 8.4. The typical codec includes EZW [12], SPIHT [13] or JBIG [14]. Among them, EZW and SPIHT produce a distortion-progressive bitstream based on the zero-tree's downward dependencies in multilevel discrete wavelet transformation (DWT), and JBIG generates the progressive bitstream based on bit-plane coding of DCT blocks.

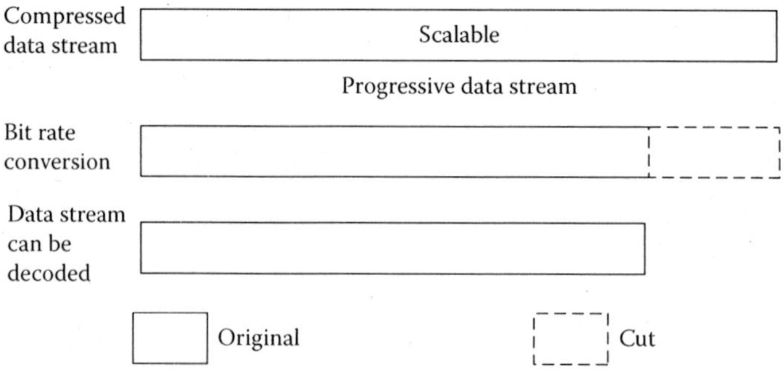

Figure 8.4   The data stream in progressive coding.

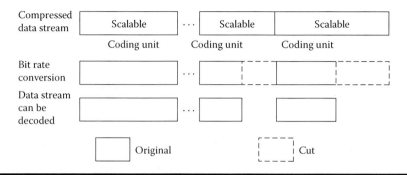

Figure 8.5   The data stream in scalable coding.

8.2.4   Scalable Coding

The fourth method, scalable coding, compresses media data into a scalable bitstream that can be reordered. Generally, the bitstream is composed of small coding units, and has various scalabilities, that is, SNR scalability, spatial scalability, and temporal scalability. After reordering the coding units, the bitstream can support the corresponding scalability. Additionally, the coding unit is scalable and can be truncated directly, as shown in Figure 8.5. The typical codecs are JPEG2000 [15] and MPEG4 SVC. JPEG2000 generates the reordered bitstream with both resolution scalability and SNR scalability. In this codec, the image is transformed with octave-band decomposition, divided into blocks, and quantized and encoded with bit-plane encoding. The smallest coding unit is the coding pass that is composed of a variable number of bits. MPEG4 SVC (Scalable Video Coding) [16], as a scalable extension of MPEG4 H.264/AVC [17], generates a bitstream with various scalabilities, including spatial scalability, temporal scalability, and SNR scalability. Generally, the bitstream is packaged into Network Abstract Layer (NAL) Units, and the scalability information is stored in the Supplemental Enhancement Information (SEI) messages.

8.3 Scalable Encryption Algorithms

According to various scalable coding methods, scalable encryption schemes should be designed in order to maintain the data stream's scalability as much as possible. The encryption scheme suitable for each scalable coding method is presented below. In each encryption scheme, two issues will be solved. The first is the kind of cipher used to encrypt each layer or coding unit. The second is, what is the relation between the keys of different layers or coding units?

8.3.1 Scalable Encryption for Layered Coding

In layered coding, the base layer, middle layer or enhancement layer is either entirely decoded or not decoded. Thus, each layer can be encrypted by a stream

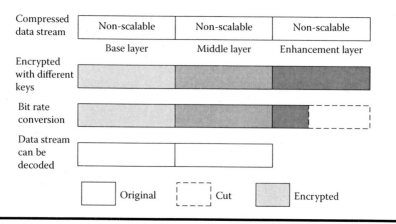

Figure 8.6 Bit rate conversion with layered encryption.

cipher or block cipher. Additionally, because the data layers cannot be reordered, the keys of the various layers may be either the same or different, and they can be generated one by one. The bit rate conversion process with scalable encryption is shown in Figure 8.6. As can be seen, the original data stream's scalability can be kept unchanged.

8.3.2 Scalable Encryption for Layered and Progressive Coding

In layered and progressive coding, the base layer is non-scalable, while the enhancement layer is scalable. Thus, the base layer can be encrypted by either a stream cipher or a block cipher. Considering that the enhancement layer may be truncated in bit-level granularity, it can only be encrypted by a stream cipher because a block cipher will cause some losses to the decrypted data. Here, the stream cipher may also be constructed on a block cipher, such as the counter (CTR) mode [18]. Additionally, the keys of the base layer and the enhancement layer may be the same or different. Thus, the bit rate conversion process with scalable encryption, as shown in Figure 8.7, keeps the original data stream's scalability unchanged.

8.3.3 Scalable Encryption for Progressive Coding

In progressive coding, the compressed data stream is scalable. Considering that the data stream may be truncated in bit-level granularity, it can only be encrypted by a stream cipher. The reason and method are the same as for the enhancement layer in layered and progressive coding. Additionally, one key is used to encrypt the whole stream. Thus, the bit rate conversion process with scalable encryption, as shown in Figure 8.8, keeps the original data stream's scalability unchanged.

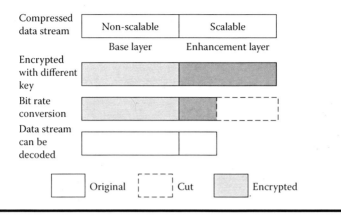

Figure 8.7 Bit rate conversion with layered and progressive encryption.

8.3.4 Scalable Encryption for Scalable Coding

In scalable coding, the compressed data stream is composed of small coding units, and they are scalable and can be reordered. To keep the bit-level granularity, each coding unit should be encrypted with a stream cipher, which is similar to the encryption of progressive coding. The data stream in scalable coding contains many coding units that may be removed. Thus, the keys of the coding units should be carefully decided. In the first case, different coding units are encrypted with the same key. Thus, it is easy to synchronize the key even after bit rate conversion. However, considering that the coding unit may be composed of few bits, encryption based on the same key is not secure enough. In the second case, different coding units are encrypted with different keys. The security will be improved greatly. But the difficulty is to synchronize the keys even after coding unit removal or reordering. To achieve this, some ancillary information is required on the order of

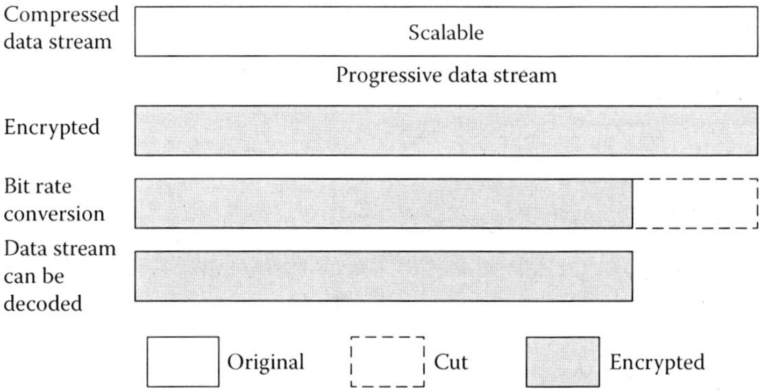

Figure 8.8 Bit rate conversion with progressive encryption.

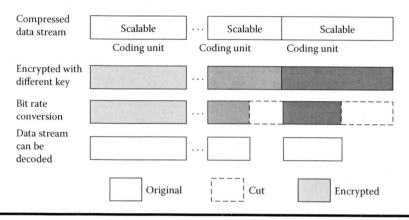

Figure 8.9   Bit rate conversion with scalable encryption.

the coding units. Of course, the ancillary information should be transmitted in a secure manner. Suppose the issue of key synchronization is solved, the bit rate conversion process with scalable encryption, as shown in Figure 8.9, keeps the original data stream's scalability unchanged.

8.4 Performance Comparison

According to the previous analysis, the ciphers and key synchronization means suitable for different scalable coding methods are compared and listed in Table 8.2. As can be seen, for the progressive or scalable stream, a stream cipher is more suitable than a block cipher because of the bit-level granularity. For the data stream that may be reordered, the contained coding units can be encrypted by different keys.

Table 8.2   Encryption Schemes for Different Scalable Coding Methods

Scalable Coding Method	Suitable Cipher	Key Synchronization
Layered coding	Block cipher or stream cipher	Same key or different key
Layered and progressive coding	Base layer: block cipher or stream cipher Enhancement layer: stream cipher	Same key or different key
Progressive coding	Stream cipher	Same key
Scalable coding	Stream cipher	Same key: low security Different key: ancillary information is required

In this case, information on the order of the coding units is required to realize key synchronization.

Considering the scalable data stream's properties, the encryption operation is often applied to the compressed data stream. Thus, the compression efficiency is unchanged. The security depends on the adopted ciphers. The encryption efficiency can be improved by encrypting only some sensitive layers or coding units. For example, in layered coding, only the base layer and middle layer are encrypted, while leaving the enhancement layer unchanged. For details, see the chapter on partial encryption.

8.5 Summary

In this chapter, scalable encryption is defined and introduced, and some typical scalable encryption algorithms suitable for different scalable coding methods are presented. The encryption algorithms include layered encryption, layered and progressive encryption, progressive encryption and scalable encryption. They are suitable for data streams encoded by layered coding, layered and progressive coding, progressive coding and scalable coding, respectively. For each encryption algorithm, the suitable ciphers and key synchronization method are proposed. Scalable encryption can also be regarded as a kind of partial encryption. Its typical application is secure media transcoding.

References

[1] W. Li. 2001. Overview of Fine Granularity Scalability in MPEG-4 Video Standard. *IEEE Transactions on Circuits and Systems for Video Technology* 11(3): 301–317.

[2] N. Conci. 2007. Initial solution for scalable and adaptive coding, TPR of multistandard integrated network convergence for global mobile and broadcast technologies (MING-7). http://ming-t.informatik.uni-hamburg.de/images/promotional/ming-t_pu_deliverable_3_3_v1_0.pdf

[3] Y. Chang, R. Han, C. Li, and J. R. Smith. 2002. Secure transcoding of Internet content. In *Proceedings International Workshop on Intelligent Multimedia Computing and Networking (IMMCN)* Durham, NC, March 8–12, 940–943.

[4] T. Kunkelmann, and U. Horn. 1998. Partial video encryption based on scalable coding. Paper presented at the 5th International Workshop on Systems, Signals and Image Processing (IWSSIP '98), June.

[5] A. S. Tosun, and W.-C. Feng. 2000. Efficient multi-layer coding and encryption of MPEG video streams. In *Proceedings IEEE International Conference on Multimedia and Expo*, Vol. 1, 119–122.

[6] S. J. Wee, and J. G. Apostolopoulos. 2003. Secure scalable streaming and secure transcoding with JPEG-2000. *IEEE International Conference on Image Processing*, 14–17 September, Barcelona, Spain, Vol. 1, I-205–208.

[7] C. Yuan, B. Zhu, Y. Wang, S. Li, and Y. Zhong. 2003. Efficient and fully scalable encryption for MPEG-4 FGS. In *Proceedings IEEE International Symposium on Circuits and Systems*, May 25–28, Bangkok, Thailand, Vol. 2: 620–623.

[8] B. Zhu, C. Yuan, Y. Wang, and S. Li. 2005. Scalable protection for MPEG-4 Fine Granularity scalability. *IEEE Transactions on Multimedia* 7(2): 222–233.

[9] ISO/MPEG-2. ISO 13818-2: Coding of moving pictures and associated audio, 1994.

[10] ITU-T Recommendation H.263-Video Coding for Low Bit Rate Communication.

[11] J. Ridge, Y. Bao, M. Karczewicz, and X. Wang. 2005. Cyclical block coding for FGS. ISO/IEC JTC1/SC29/WG11, M11509. *Proposal of MPEG4 Standard.*

[12] J. M. Shapiro. 1992. Embedded image coding using zerotrees of wavelet coefficients. *IEEE Transactions on Signal Processing* 41: 657–660.

[13] A. Said. 1996. A new fast and efficient image codec based on set partitioning in hierarchical trees. *IEEE Transactions on Circuits and Systems for Video Technology* 6: 243–250.

[14] JBIG lossless image compression standard, ISO/IEC standard 11544 and ITU-T recommendation T.82.

[15] D. Taubman, E. Ordentlich, M. Weinberger, and G. Seroussi. 2001. *Embedded Block Coding in JPEG2000.* Loveland, CO: Hewlett-Packard Company.

[16] J. Reichel, M. Wien, and H. Schwarz. Scalable Video Model 3. ISO/IEC JTC1/SC29/WG11, N6716, 2004.

[17] H.264/MPEG4 Part 10 (ISO/IEC 14496-10): Advanced Video Coding (AVC), ITU-T H.264 standard.

[18] R. A. Mollin. 2006. *An Introduction to Cryptography.* Boca Raton, FL: CRC Press.

Chapter 9

Commutative Watermarking and Encryption

9.1 Definition of Commutative Watermarking and Encryption

As is known, encryption is used to protect the confidentiality of media content, and watermarking [1, 2] can be used to protect the copyright of media content. The watermarking technique embeds copyright information into media content by modifying the media pixels slightly. The embedded information can be detected or extracted from media content and used to tell the ownership.

Because media encryption and media watermarking serve different functions, they can be combined to protect both confidentiality and ownership/identity. Generally, it is implemented in two steps [3, 4]. First, the media data is watermarked. Second, the watermarked media data is encrypted. In this case, media data must be decrypted before the watermark can be detected or another watermark can be embedded. That is, the encryption operation and watermarking operation cannot be commutated. In some applications, if they are commutative, some computing cost will be saved. For example, the watermark can be embedded into the encrypted media data directly, or it can be extracted from the decrypted media data. Additionally, the direct operation in the encrypted domain can be supported, which is suitable for some secure applications. For example, the watermark can be embedded

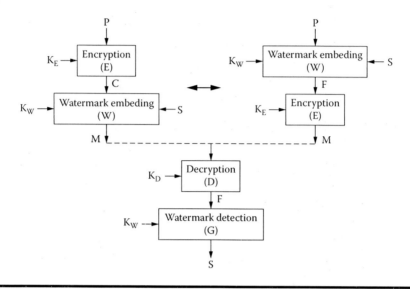

Figure 9.1   Architecture of commutative watermarking and encryption.

into the encrypted media data directly without knowing the decryption key, which avoids the leakage of media content.

The concept of commutative watermarking and encryption (CWE) was first reported in [5]. It means that multimedia content can either be first watermarked then encrypted or first encrypted then watermarked. The two processes are equivalent, as shown in Figure 9.1. Let P be the original media content, C the encrypted media content, M the watermarked media content, E() the encryption algorithm, W() the watermark embedding algorithm, K_E the encryption key, K_W the watermark key and S the watermark. For the commutative encryption and watermarking operations, the following condition should be satisfied.

$$W(E(P, K_E), W, K_W) = E(W(P, S, K_W), K_E) = M.$$

Furthermore, set D() the decryption algorithm, G() the watermark detection algorithm, and K_D the decryption key. Thus, the watermark can be detected according to

$$G(D(M, K_D), K_W) = G(D(E(W(P, S, K_W), K_E), K_D), K_W)$$

$$= G(W(P, S, K_W), K_W) = S.$$

As can be seen, if encryption and watermarking operations are commutative, the order of encryption or watermarking can be changed, which does not affect correct decryption or watermark detection.

However, the difficulty is to find the commutative encryption or watermarking operations. In [5], commutative operations based on exponential operation are proposed, which satisfy the following condition.

$$\left(P^{K_E}\right)^{K_W} = P^{K_E K_W} = \left(P^{K_W}\right)^{K_E}.$$

There exists the encryption algorithm based on an exponential operation, for example, RSA cipher [6]. However, it is still difficult to find the watermarking algorithm based on an exponential operation.

To date, there are two practical solutions. The first is based on partial encryption, and the other is based on homomorphic encryption and watermarking operations. They are presented in detail below.

9.2 CWE Based on Partial Encryption

The first solution makes use of partial encryption to construct the CWE scheme. In this kind of scheme [7, 8], media data is partitioned into two parts, one is encrypted and the other is watermarked. For the encrypted part to be independent of the watermarked part, the properties of the encryption algorithm and the watermarking algorithm are kept unchanged. However, the disadvantage is that only a part of the media data is encrypted, which causes some loss in security.

9.2.1 The General Architecture

The proposed scheme is shown in Figure 9.2. Let the original media P be composed of two parts, X_0 and X_1. X_0 is the perception-significant part, changes to which often cause the media data to be unintelligible, whereas human eyes are not as sensitive to X_1. In partial encryption, only the perception-significant part X_0 is encrypted into X'_0, while X_1 is left unchanged. In watermarking, because of the redundancy in X_1, information S is embedded into X_1 imperceptibly, which produces the data X'_1. Thus, X'_0 and X'_1 form the encrypted and marked data M. The encryption and watermarking processes are defined as

$$\begin{cases} X'_0 = \mathrm{E}(X_0, K_E) \\ X'_1 = \mathrm{W}(X_1, S, K_W) \end{cases}.$$

In decryption and watermark detection, the followed operations are done.

$$\begin{cases} X_0 = \mathrm{D}(X'_0, K_D) \\ S = \mathrm{G}(X'_1, K_W) \end{cases}.$$

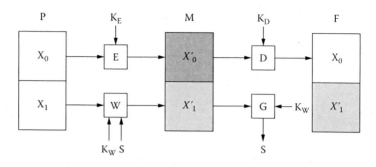

Figure 9.2 Architecture of the CWE scheme based on partial encryption.

Because the encryption part X_0 and the watermarking part X_1 are independent from each other, the encryption algorithm's property and watermarking algorithm's property are both unchanged. However, the security is decreased because the watermarking part is not encrypted.

9.2.2 The Improved Scheme

The scheme based on partial encryption can be improved by introducing a random partitioning operation. That is, the media P is partitioned into two parts, X_0 and X_1, under the control of a key. As shown in Figure 9.3, whether a pixel in media P belongs to X_0 or X_1 is decided by the key bit. Thus, besides the encryption key K_E and watermarking key K_W, the partitioning key K is also required by a user who can decrypt the media data or extract the watermark correctly. However, under the condition of known-plaintext attack or select-plaintext attack, the

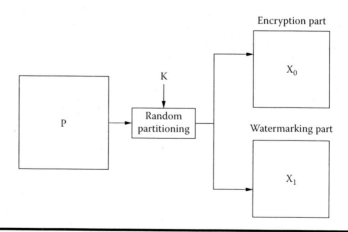

Figure 9.3 The improved scheme based on random partitioning.

Table 9.1   Media Data Partitioning

Media	Encryption Part X_0	Watermarking Part X_1
Raw image	Significant bits, foreground	Nonsignificant bits, background
Raw audio	Sensitive segments	Other segments
Raw video	Foregrounds, sensitive frames	Backgrounds, other frames
Compressed image, audio or video	Sensitive parameters	Other parameters

partitioning mode can be broken, unless different media content is partitioned with different keys.

9.2.3   Selection of the Encryption Part and Watermarking Part

The selection of the encryption part and watermarking part should consider the perceptual security and the robustness of the watermark. According to partial encryption, the data part that is significant to the media's intelligibility is preferred to be encrypted. The watermarking part should be robust against some acceptable operations, such as recompression or slight noise. Considering the properties of images, audios or videos, we can partition them into two parts according to Table 9.1. Taking a raw image, for example, the significant bits in image pixels can be regarded as the significant part and the nonsignificant ones as the second part. For raw video, the first part may be composed of foregrounds or sensitive frames, while the second part may be composed of backgrounds or other frames. For compressed media data, the first part may include some sensitive parameters, while the second part is composed of other parameters.

Taking compressed video for example, the MEPG4 AVC/H.264 [9] video is encrypted and watermarked with the partial encryption based CWE scheme. As presented in [8], such parameters as inter/intra-prediction mode (IPM), motion vector difference (MVD) and residue coefficient sign are encrypted, while the amplitude of DC or AC is watermarked with quantization embedding. That is, the watermarking operation only changes the bit-planes of discrete cosine transform (DCT) coefficients. To save time, the selected parameters are encrypted partially. To keep them robust and imperceptible, the coefficients are selected adaptively according to macroblock type. Figure 9.4 shows encrypted video that is unintelligible. The quality of the watermarked video is shown in Figure 9.5. Here, Salesman is QCIF, Mobile is CIF, and they are both encoded at 3 Mbps. As can be seen, when the quantization step (QP) is no bigger than 36, the video quality is acceptable. Additionally, the watermarking and encryption operations do not affect each other, that is, the commutation property is satisfied.

Original   Encrypted

Figure 9.4   **Video encryption based on MPEG4 AVC/H.264.**

9.3 CWE Based on Homomorphic Operations

There exist some homomorphic [6] encryption and watermarking operations, which can be used to design the CWE scheme. In the following content, bitwise XOR operation and module addition operations will be investigated.

9.3.1 The CWE Scheme Based on XOR Operation

Considering that XOR operation is often used in either encryption [6] or watermarking [10], the commutative scheme based on XOR operation can be constructed. Taking XOR operation for example, the encryption operation is

$$C = E(P, K_E) = P \oplus K_E.$$

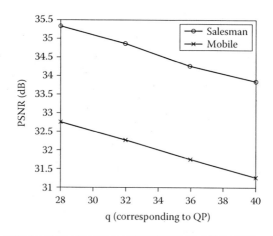

Figure 9.5   **Quality of the watermarked video.**

Here, $\oplus$ is the bitwise XOR, P is the pixel in the original media content, and K_E is the pixel in the key stream that can be generated from a pseudorandom sequence generator [6, 10]. Generally, P is of L-bit, K_E is of Q_0-bit *(0 < $Q_0 \leq L$)*. Q_0 is in relation with the security of the encryption algorithm. The watermarking embedding operation is

$$M = W(C, K_W) = C \oplus K_W.$$

Here, C is the pixel in the encrypted media content and K_W is the pixel in the watermark sequence that can be generated from a pseudorandom sequence generator. Generally, C is of L-bit, K_W is of Q_1-bit *(0 < $Q_1 \leq L$)*. Generally, Q_1 is small enough to maintain the imperceptibility of the watermark. The decryption operation is

$$F = D(M, K_D) = M \oplus K_D.$$

Here, M is the pixel in the watermarked media content, and K_D is the pixel in the decryption key stream that can be generated from a pseudorandom sequence generator. Generally, K_D is equal to K_E. Thus, we get

$$W(E(P, K_E), K_W) = P \oplus K_E \oplus K_W = P \oplus K_W \oplus K_E = E(W(P, K_W), K_E)$$

and

$$D(W(E(P, K_E), K_W), K_D) = P \oplus K_E \oplus K_W \oplus K_D$$
$$= P \oplus K_W \oplus K_E \oplus K_D = P \oplus K_W = W(P, K_W).$$

Here, $K_E = K_D$. As can be seen, when E() and W() are both XOR operation, they are commutative. From the received media copy F, the watermark S can be detected according to the following method.

$$S = \begin{cases} K_W, & \text{if} < F \oplus P, K_W \gg T \\ \neq K_W, & \text{otherwise} \end{cases}.$$

Here, <A,B> denotes the correlation value between A and B (A and B are two samples), and T is the threshold used to determine the existence of the watermark.

Taking raw image encryption, for example, the image pixels are encrypted and watermarked by bitwise XOR operation. The encrypted image, watermarked image and decrypted image are shown in Figure 9.6. Here, $Q_0 = 8$ and $Q_1 = 3$. As can be seen, the encryption and watermarking operation are commutative.

Figure 9.6 **The images generated by the XOR-based CWE scheme.**

9.3.2 The CWE Scheme Based on Module Addition

In another case, for additive watermarking and encryption based on random modulation, the commutative watermarking and encryption scheme can also be constructed. Similarly, the encryption operation and watermark embedding are defined as

$$\begin{cases} C = E(P, K_E) = (P' + K_E) \bmod L \\ M = W(C, K_W) = (C + K_W) \bmod L \end{cases}.$$

Here, P $(0 \le P < L)$ is the pixel in the original media content, C $(0 \le C < L)$ is the pixel in the encrypted media content, K_E $(0 \le K_E < L)$ is the pixel in the encryption sequence, K_W $(-A \le K_W < A)$ is the pixel in the watermark sequence, and P' $(A \le P <$

$L - A$) is the preprocessed pixel in the original media content. As can be seen, the original pixel P is preprocessed by shifting to the range of $[A, L - A - 1]$.

The decryption operation is done according to

$$F = D(M, K_E) = (M - K_E) \bmod L$$

$$= (P' + K_E + K_W - K_E) \bmod L = (P' + K_W) \bmod L = P' + K_W.$$

Then, the watermark is detected by

$$S = \begin{cases} K_W, & if < F, K_W \gg T \\ \neq K_W, & otherwise \end{cases}.$$

Here, T is the threshold used to determine the existence of the watermark.

Taking MPEG2 video encryption, for example, the DC coefficient of each luminance DCT block is encrypted and watermarked by modulation operation. The encrypted video, watermarked video and decrypted video are shown in Figure 9.7. Here, $L = 1024$ and $A = 32$. As can be seen, the encryption and watermarking operation are commutative. Of course, only encrypting DC coefficients is not secure enough, which can be improved by encrypting chrominance DC coefficients, AC coefficients' signs and motion vectors' signs with other algorithms. Figure 9.7(f) shows the video encrypted by the improved scheme. Here, except luminance DC, the selected parameters are encrypted with AES. As is apparent, the watermarking and encryption remain commutative.

9.4 Performance Comparison

The partial encryption CWE schemes and the homomorphic operation CWE schemes have different properties. Here, two aspects are considered, the security of encryption and the robustness of the watermark, as shown in Table 9.2. For the partial encryption scheme, the watermark's robustness can be obtained by selecting a robust watermarking part. The media content's properties can be used to embed the watermark adaptively. In the homomorphic CWE scheme, the watermark's robustness is limited by the encryption operation. For example, the pixels' statistical properties are changed by the encryption operation, which cannot be used to embed the watermark adaptively. With respect to the content security, the partial encryption scheme depends on the partial encryption algorithm adopted, while for the homomorphic scheme, the security of the content is based on the stream cipher adopted. Thus, if the partial encryption can satisfy the required security, the partial encryption CWE scheme is preferred for practical applications.

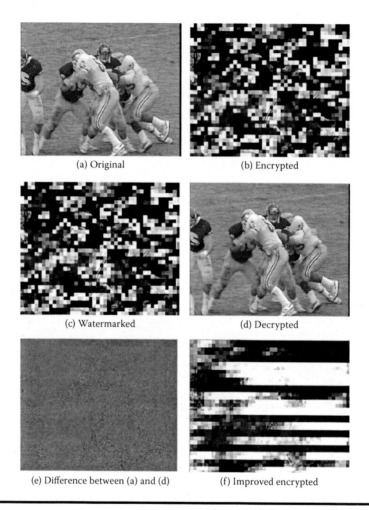

(a) Original

(b) Encrypted

(c) Watermarked

(d) Decrypted

(e) Difference between (a) and (d)

(f) Improved encrypted

Figure 9.7   The MPEG2 video generated by the module addition CWE scheme.

Table 9.2   Performance Comparison of Different CWE Schemes

CWE Method	Content Security	Robustness of the Watermark
CWE scheme based on partial encryption	Based on partial encryption	Robust
Improved partial encryption-based CWE	Based on partial encryption and stream cipher	Relatively robust
CWE scheme based on homomorphic operations	Based on stream cipher	Not robust

9.5 Summary

In this chapter, commutative watermarking and encryption (CWE) are defined and introduced, and some typical CWE schemes are presented. Typical schemes include the partial encryption scheme and the homomorphic scheme. The former easily provides high robustness of the watermark, while the latter one provides higher security. Note that, for homomorphic CWE, homomorphic encryption and watermarking operations are required. However, to date, the known homomorphic operations are only limited to stream ciphers. Thus, the homomorphic operations in block ciphers are anticipated for this method to find wide application.

References

[1] P. Moulin, and R. Koetter. 2005. Data-hiding codes. *IEEE Proceedings* 93(12): 2083–2126.

[2] J. A. Bloom, I. J. Cox, T. Kalker, J. P. Linnartz, M. L. Miller, and C. B. Traw. 1999. Copy protection for digital video. *IEEE Proceedings* (Special Issue on Identification and Protection of Multimedia Information) 87(7): 1267–1276.

[3] T. Wu, and S. Wu. 1997. Selective encryption and watermarking of MPEG video. Paper presented at International Conference on Image Science, Systems and Technology, CISST'97, Los Angeles, CA, February.

[4] D. Simitopoulos, N. Zissis, P. Georgiadis, V. Emmanouilidis, and M. G. Strintzis. 2003. Encryption and watermarking for the secure distribution of copyrighted MPEG video on DVD. *ACM Multimedia Systems Journal* (Special Issue on Multimedia Security) 9(3): 217–227.

[5] European Network of Excellence in Cryptology. 2005. First summary report on hybrid systems. TR: IST-2002-507932, ECRYPT. http://www.ecrypt. eu.org/documents/

[6] A. R. Mollin. 2006. *An Introduction to Cryptography*. Boca Raton, FL: CRC Press.

[7] S. Lian, Z. Liu, Z. Ren, and H. Wang. 2006. Commutative watermarking and encryption for media data. *International Journal of Optical Engineering* 45(8): 0805101–0805103.

[8] S. Lian, Z. Liu, Z. Ren, and H. Wang. 2007. Commutative encryption and watermarking in compressed video data. *IEEE Circuits and Systems for Video Technology* 17(6): 774–778.

[9] H.264/MPEG4 Part 10 (ISO/IEC 14496-10): Advanced Video Coding (AVC), ITU-T H.264 standard.

[10] M. U. Celik, G. Sharma, A. M. Tekalp, and E. Saber. 2005. Lossless generalized-LSB data embedding. *IEEE Transactions on Image Processing* 14(2): 253–266.

Chapter 10

Joint Fingerprint Embedding and Decryption

10.1 Definition of Joint Fingerprint Embedding and Decryption

Digital fingerprinting [1–4] is a technology used to trace illegal distributors, which embeds users' identification information imperceptibly into media content. The user information, such as an ID code, is embedded into the media content by watermarking techniques [5, 6]. If the customer redistributes his media copy to other unauthorized customers, e.g., by putting it on the Internet, the ID code can be extracted from the media content and identify illegal distributors.

In secure media distribution, media content is often encrypted before transmission on the sender side. In order to trace illegal distributors, the fingerprint information is embedded on the receiver side, which is suitable for applications with a large number of customers, such as video-on-demand or digital TV. However, if the decryption operation and fingerprint embedding operation are independent, as shown in Figure 10.1, the plain-content may be leaked out from the gap between decryption and fingerprint embedding. The decryption and fingerprint embedding process is defined as

$$M = W(D(C, K_D), F),$$

where C, F, M, $D()$, and $W()$ are the cipher-media, fingerprint information, fingerprinted media, decryption function and fingerprint embedding, respectively.

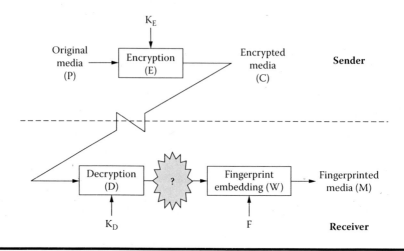

Figure 10.1   General architecture of secure media distribution.

To solve the problem of content leakage, joint fingerprint embedding and decryption scheme (JFD) [7] have been reported, as shown in Figure 10.2, which combines decryption operation and fingerprint embedding and avoids the content leakage. The JFD operation is defined as

$$M = J(C, K_D, F),$$

where C, F, M, and $J()$ are the cipher-media, fingerprint information, fingerprinted media and JFD function, respectively.

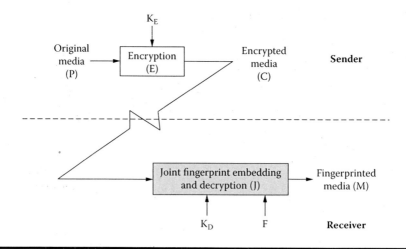

Figure 10.2   Secure media distribution based on joint fingerprint embedding and decryption (JFD).

As can be seen, the JFD operation $J()$ can be regarded as the combination of $W()$ and $D()$. Several JFD schemes have been reported. They are introduced and analyzed below.

10.2 Typical JFD Schemes

Three properties of a JFD scheme need to be investigated, the security of encryption, the robustness and imperceptibility of the fingerprint, and the security of the fingerprint. Here, the fingerprint's security denotes the ability to survive collusion attacks [8]. In a collusion attack, several customers combine their media copies together by such operations as averaging, min-max selection, Linear Combination Collusion Attack (LCCA) [9], etc.

10.2.1 The JFD Scheme Based on Broadcast Encryption

The JFD scheme based on broadcast encryption [10] is shown in Figure 10.3. On the sender side, the media stream is partitioned into segments, each segment is watermarked into two copies, and thus, two media streams are produced. In the

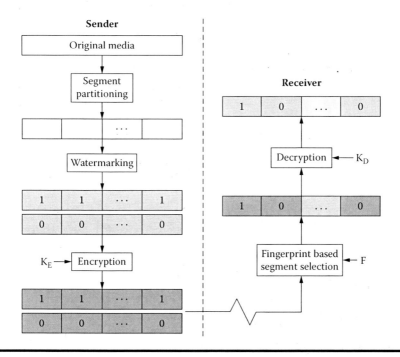

Figure 10.3 The joint fingerprint embedding and decryption (JFD) scheme based on broadcast encryption.

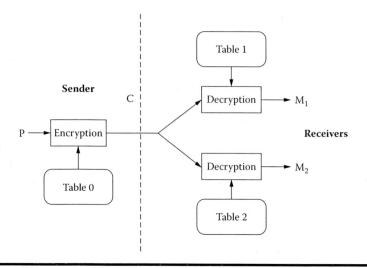

Figure 10.4   The joint fingerprint embedding and decryption (JFD) scheme based on codeword modification.

first media stream, the information bit "1" is embedded into each segment. In the second media stream, the information "0" is embedded into each segment. Then, the two media streams are encrypted. Here, each segment in either stream is encrypted with a different key generated from K_E. On the receiver side, the media stream corresponding to the fingerprint information F is received. Here, for the ith segment, whether "1" segment is received or "0" segment is received is determined by the fingerprint bit F_i. Then, the key sequence is produced by K_D, and is used to decrypt the received media stream. In this scheme, the segment can be encrypted with a strong cipher, such as AES, 3DES and so on. Thus, high security can be obtained. The disadvantage is that double streams need to be transmitted, which doubles the transmission cost.

10.2.2   The JFD Scheme Based on Codeword Modification

The codeword modification method [11] is shown in Figure 10.4. On the sender side, the media content P is encrypted into C with a secure table. On the receiver side, different users employ different tables to decrypt the media content into separate copies. The fingerprint information is determined by the decryption table. Here, there are slight differences between the three tables, that is, Table 0, Table 1, and Table 2. For example, for the corresponding codewords in Table 1 and Table 2, there is only one bit difference. Generally, the bit difference happens in the least significant bit-plane, which maintains the imperceptibility of the watermark. It was reported that the scheme is time efficient and secure against cryptographic attacks.

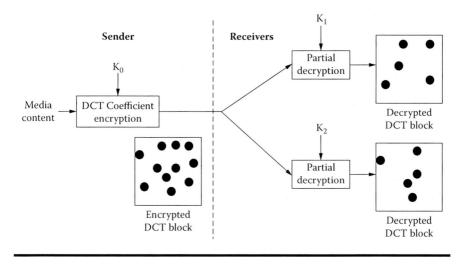

Figure 10. 5 The joint fingerprint embedding and decryption (JFD) scheme based on partial decryption.

However, for different customers, different key tables should be transmitted, which cost bandwidth. Additionally, the least significant bits are not robust to signal processing, such as recompression, additive noise, filtering, etc.

10.2.3 The JFD Scheme Based on Partial Decryption

The JFD scheme based on partial decryption [12] is shown in Figure 10.5. On the sender side, the discrete cosine transform (DCT) coefficients of the media content are all encrypted: On the receiver side, some of the coefficients are decrypted, while some are left undecrypted. By detecting the position of the undecrypted coefficients, the identification information can be explored. Thus, the difference between the decryption keys determines the difference between the undecrypted coefficients. The scheme is robust to some operations, while the imperceptibility cannot be confirmed. Because only the DCT coefficients' signs are encrypted, the encrypted media content is not secure in perception. Additionally, the security against collusion attacks cannot be confirmed.

10.2.4 The JFD Scheme Based on Index Modification

The JFD scheme based on index modification [13] is shown in Figure 10.6. On the sender side, the Huffman code is encrypted by index encryption. On the receiver side, the Huffman code is decrypted by index decryption together with index modification. The index modification depends on both the encrypted Huffman code and the fingerprint information *F*. Thus, different fingerprints will produce different

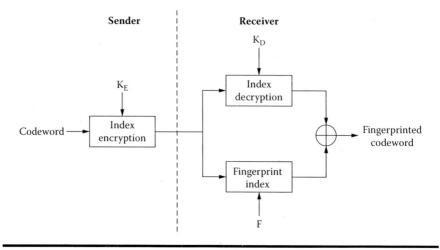

Figure 10.6 **The joint fingerprint embedding and decryption (JFD) scheme based on index modification.**

Huffman codes. This scheme is designed for MPEG data stream distribution. It provides security against some cryptographic attacks, while the robustness against some signal processing operations (recompression, adding noise, filtering, etc.) cannot be confirmed.

10.2.5 The JFD Scheme Based on Homomorphic Operations

The JFD scheme based on homomorphic operations [14] is shown in Figure 10.7. On the sender side, the media P is modulated by the random sequence R, and it is encrypted into $C = P + R$. On the receiver side, each customer gets the key sequence $R - \alpha F$ from

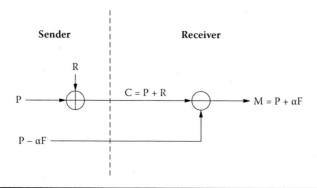

Figure 10.7 **The joint fingerprint embedding and decryption (JFD) scheme based on homomorphic operations.**

Table 10.1 Encryption Schemes for Different Scalable Coding Methods

JFD Scheme	Security	Robustness	Transmission Cost
JFD based on broadcast encryption	SL0	Robust	Double media streams + key sequence
JFD based on codeword modification	SL0	Not robust	Key table
JFD based on partial decryption	SL3	Robust	Key hash
JFD based on index modification	SL0	Not robust	Fingerprint table
JFD based on homomorphic operations	SL2	Robust	Key sequence

the sender, and uses the sequence to demodulate media content C and produce the fingerprinted media $M = P + \alpha F$. Because different customers own different F, each will obtain a different media copy. As can be seen, the scheme is constructed on the addition operation that is used in both modulation encryption and additive watermarking. The scheme is robust against signal processing, which benefits from the additive watermarking algorithms, while security against cryptographic attacks cannot be confirmed. Additionally, the transmission of the key stream costs much time and space.

10.3 Performance Comparison

According to the previous analysis, the existing JFD schemes have different properties in such aspects as security, robustness and transmission cost. The comparative results are shown in Table 10.1. Generally, all of them have some disadvantages. For example, the JFD scheme based on broadcast encryption is secure against cryptographic attacks and robust against media processing, but it has twice the transmission size. The schemes based on codeword modification and index modification are time efficient but not robust against media processing. That based on partial decryption is not secure in perception. Homomorphic operations are robust against media processing but not secure against cryptographic attacks. Thus, the trade-off between the various properties needs to be investigated.

10.4 Summary

In this chapter, joint fingerprint embedding and decryption (JFD) schemes are introduced and analyzed. They are constructed based on broadcast encryption, codeword modification, partial decryption, index modification, and homomorphic encryption, respectively. Their architectures and performances are investigated and compared. With respect to the properties of security, robustness and transmission cost, they all need to be improved. This technique is still young and more research is expected. However, from another viewpoint, the JFD scheme is designed for the untrusted user side. If the decryption and fingerprinting can be implemented in a trusted computing platform, then a JFD scheme is no longer needed.

References

[1] A. Herrigel, J. Oruanaidh, H. Petersen, S. Pereira, and T. Pun. 1998. Secure copyright protection techniques for digital images. In Proceedings Second Information Hiding Workshop (IHW). Lecture Notes in Computer Science, 1525, 169–190.

[2] D. Boneh, and J. Shaw. 1998. Collusion-secure fingerprinting for digital data. *IEEE Transactions on Information Theory* 44: 1897–1905.

[3] Z. J. Wang, M. Wu, W. Trappe, and K. J. R. Liu. 2005. Group-oriented fingerprinting for multimedia forensics. Preprint.

[4] W. Trappe, M. Wu, Z. J. Wang, and K. J. R. Liu. 2003. Anti-collusion fingerprinting for multimedia. *IEEE Transactions on Signal Processing* 51: 1069–1087.

[5] P. Moulin and R. Koetter. 2005. Data-hiding codes. *IEEE Proceedings* 93(12): 2083–2126.

[6] J. A. Bloom, I. J. Cox, T. Kalker, J. P. Linnartz, M. L. Miller, and C. B. Traw. 1999. Copy protection for digital video. *IEEE Proceedings* (Special Issue on Identification and Protection of Multimedia Information) 87(7): 1267–1276.

[7] S. Lian, Z. Liu, Z. Ren, and H. Wang. 2007. Joint fingerprint embedding and decryption for video distribution. Paper presented at IEEE International Conference on Multimedia and Expo (ICME2007), Beijing, China.

[8] M. Wu, W. Trappe, Z. J. Wang, and R. Liu. 2004 (March). Collusion-resistant fingerprinting for multimedia. *IEEE Signal Processing Magazine,* 15–27.

[9] Y. Wu. 2005. Linear combination collusion attack and its application on an anti-collusion fingerprinting. In *Proceedings IEEE International Conference on Acoustics, Speech, and Signal Processing,* (ICASSP '05), March 18–23, Vol. 2, 13–16.

[10] R. Parnes, and R. Parviainen. 2001. Large scale distributed watermarking of multicast media through encryption. In *Proceedings IFIP International Conference on Communications and Multimedia Security Issues of the New Century,* 17.

[11] R. Anderson, and C. Manifavas. 1997. Chameleon: A New Kind of Stream Cipher. Lecture Notes in Computer Science, Fast Software Encryption, 107–113.

[12] D. Kundur, and K. Karthik. 2004. Video fingerprinting and encryption principles for digital rights management. *IEEE Proceedings* 92(6): 918–932.

[13] S. Lian, Z. Liu, Z. Ren, and H. Wang. 2006. Secure distribution scheme for compressed data streams. Paper presented at 2006 IEEE Conference on Image Processing (ICIP 2006), Atlanta, GA, October 8-11, 1953–1956.

[14] A. N. Lemma, S. Katzenbeisser, M. U. Celik, and M. V. Veen. 2006. Secure watermark embedding through partial encryption. In *Proceedings International Workshop on Digital Watermarking (IWDW 2006).* Lecture Notes in Computer Science, 4283, 433–445.

Chapter 11

Typical Attacks on Multimedia Encryption

11.1 Security Requirement of Multimedia Encryption

As mentioned in the chapter on performance requirements, multimedia encryption has special security requirements. Generally, besides cryptographic security, perceptual security is required. Additionally, for partial encryption, the selected parameters should also satisfy some principles, described in detail below.

- Cryptographic security—The encryption algorithm is secure against cryptographic attacks [1], e.g., ciphertext-only attack, known-plaintext attack, select-plaintext attack, etc. In the case of complete encryption, the traditional cipher or new cipher adopted is secure from a cryptographic viewpoint. In partial encryption, some parameters are encrypted with a cipher that is secure against cryptographic attacks. In compression-combined encryption, the encryption operation combined with compression operation is secure from a cryptographic viewpoint.
- Perceptual security—The encrypted media content is secure to human perception. For example, the encrypted image or video is unintelligible, and the encrypted audio is too chaotic to be understood. This is a special requirement of multimedia encryption. As is known, in text/binary data encryption, the encrypted content is often unintelligible. Compared with text/binary data, multimedia data often has a large redundancy. Thus, traditional encryption may lose the redundant information. For example,

using the raw image with block cipher's ECB mode will produce a cipher-image whose border information is still known. Additionally, in partial encryption, the perceptual security depends on the selected parameters. Thus, for multimedia encryption algorithms, perceptual security should be investigated.

■ Parameter security in partial encryption—In partial encryption, some parameters are encrypted with a cipher and the other parameters are left unchanged. Although the adopted cipher is strong, the security of the partial encryption may not be confirmed. It depends on the relation between the encrypted parameters and unencrypted parameters. If the encrypted parameters can be replaced by or deduced from other parameters, the plaintext will be recovered under ciphertext-only attack. Thus, the parameters should be carefully selected in partial encryption.

Some types of attack are introduced below to analyze existing multimedia encryption algorithms.

11.2 Cryptographic Attacks

Generally, the cipher should be carefully designed and analyzed before being used in practical applications. Ciphers whose security can be confirmed are classified into two types, ciphers based on iterated confusion and diffusion and ciphers based on mathematical theory. The former denotes such block ciphers as 3DES [2], AES [3], etc. Their security depends on the computing complexity benefiting from the iterated operations. The second type includes such public ciphers [4] as RSA, ECC, etc. Their security depends on the NP (nondeterministic polynomial time) problem in mathematics, such as factoring a large prime number or solving the discrete logarithm.

Recently, some new ciphers have been reported, for example, ciphers based on chaos [5, 6] or neural networks [7, 8]. Their security needs to be analyzed before we can use them. Here, taking chaotic permutation [9] for example, its security is analyzed as follows.

Taking Cat map [9] for example, in a square area $M \times M$, the Cat map is defined as

$$
\begin{pmatrix} x_{k+1} \\ y_{k+1} \end{pmatrix} = C(x_k, y_k) = \begin{pmatrix} 1 & b \\ a & ab+1 \end{pmatrix} \begin{pmatrix} x_k \\ y_k \end{pmatrix} \bmod M,
$$

where (x_k, y_k) is the position of the pixel in the original square area and satisfies $0 \le x_k, y_k < M$, k is the iteration time, a, b are the control parameters satisfying

$0 \leq a, b < M$ (a and b are integers), and (x_{k+1}, y_{k+1}) is the position of the pixel in the permuted square area.

Under ciphertext-only attack, the brute-force space is M^2. That is, the brute-force attackers need to do M^2 times of enumeration before they can recover the plaintext and parameters. Additionally, in each enumeration, each pixel should be transformed in order to judge whether the transformed image is the plain-image or not. However, under select-plaintext attack, the enumeration times can be reduced greatly. Taking $M = 16$ for example, the original 16×16 image P is set as the following gray level.

$$
P = \begin{bmatrix}
0 & 16 & 32 & 48 & 64 & 80 & 96 & 112 & 128 & 144 & 160 & 176 & 192 & 208 & 224 & 240 \\
1 & 17 & 33 & 49 & 65 & 81 & 97 & 113 & 129 & 145 & 161 & 177 & 193 & 209 & 225 & 241 \\
2 & 18 & 34 & 50 & 66 & 82 & 98 & 114 & 130 & 146 & 162 & 178 & 194 & 210 & 226 & 242 \\
3 & 19 & 35 & 51 & 67 & 83 & 99 & 115 & 131 & 147 & 163 & 179 & 195 & 211 & 227 & 243 \\
4 & 20 & 36 & 52 & 68 & 84 & 100 & 116 & 132 & 148 & 164 & 180 & 196 & 212 & 228 & 244 \\
5 & 21 & 37 & 53 & 69 & 85 & 101 & 117 & 133 & 149 & 165 & 181 & 197 & 213 & 229 & 245 \\
6 & 22 & 38 & 54 & 70 & 86 & 102 & 118 & 134 & 150 & 166 & 182 & 198 & 214 & 230 & 246 \\
7 & 23 & 39 & 55 & 71 & 87 & 103 & 119 & 135 & 151 & 167 & 183 & 199 & 215 & 231 & 247 \\
8 & 24 & 40 & 56 & 72 & 88 & 104 & 120 & 136 & 152 & 168 & 184 & 200 & 216 & 232 & 248 \\
9 & 25 & 41 & 57 & 73 & 89 & 105 & 121 & 137 & 153 & 169 & 185 & 201 & 217 & 233 & 249 \\
10 & 26 & 42 & 58 & 74 & 90 & 106 & 122 & 138 & 154 & 170 & 186 & 202 & 218 & 234 & 250 \\
11 & 27 & 43 & 59 & 75 & 91 & 107 & 123 & 139 & 155 & 171 & 187 & 203 & 219 & 235 & 251 \\
12 & 28 & 44 & 60 & 76 & 92 & 108 & 124 & 140 & 156 & 172 & 188 & 204 & 220 & 236 & 252 \\
13 & 29 & 45 & 61 & 77 & 93 & 109 & 125 & 141 & 157 & 173 & 189 & 205 & 221 & 237 & 253 \\
14 & 30 & 46 & 62 & 78 & 94 & 110 & 126 & 142 & 158 & 174 & 190 & 206 & 222 & 238 & 254 \\
15 & 31 & 47 & 63 & 79 & 95 & 111 & 127 & 143 & 159 & 175 & 191 & 207 & 223 & 239 & 255
\end{bmatrix}.
$$

Here, pixels in different positions have different gray levels. Taking $a = 2$ and $b = 15$ for example, the permuted image C is

$$
C = \begin{bmatrix}
62 & 77 & 92 & 107 & 122 & 137 & 152 & 167 & 182 & 197 & 212 & 227 & 242 & 1 & 16 & 47 \\
93 & 108 & 123 & 138 & 153 & 168 & 183 & 198 & 213 & 228 & 243 & 2 & 17 & 32 & 63 & 78 \\
124 & 139 & 154 & 169 & 184 & 199 & 214 & 229 & 244 & 3 & 18 & 33 & 48 & 79 & 94 & 109 \\
155 & 170 & 185 & 200 & 215 & 230 & 245 & 4 & 19 & 34 & 49 & 64 & 95 & 110 & 125 & 140 \\
186 & 201 & 216 & 231 & 246 & 5 & 20 & 35 & 50 & 65 & 80 & 111 & 126 & 141 & 156 & 171 \\
217 & 232 & 247 & 6 & 21 & 36 & 51 & 66 & 81 & 96 & 127 & 142 & 157 & 172 & 187 & 202 \\
248 & 7 & 22 & 37 & 52 & 67 & 82 & 97 & 112 & 143 & 158 & 173 & 188 & 203 & 218 & 233 \\
23 & 38 & 53 & 68 & 83 & 98 & 113 & 128 & 159 & 174 & 189 & 204 & 219 & 234 & 249 & 8 \\
54 & 69 & 84 & 99 & 114 & 129 & 144 & 175 & 190 & 205 & 220 & 235 & 250 & 9 & 24 & 39 \\
85 & 100 & 115 & 130 & 145 & 160 & 191 & 206 & 221 & 236 & 251 & 10 & 25 & 40 & 55 & 70 \\
116 & 131 & 146 & 161 & 176 & 207 & 222 & 237 & 252 & 11 & 26 & 41 & 56 & 71 & 86 & 101 \\
147 & 162 & 177 & 192 & 223 & 238 & 253 & 12 & 27 & 42 & 57 & 72 & 87 & 102 & 117 & 132 \\
178 & 193 & 208 & 239 & 254 & 13 & 28 & 43 & 58 & 73 & 88 & 103 & 118 & 133 & 148 & 163 \\
209 & 224 & 255 & 14 & 29 & 44 & 59 & 74 & 89 & 104 & 119 & 134 & 149 & 164 & 179 & 194 \\
240 & 15 & 30 & 45 & 60 & 75 & 90 & 105 & 120 & 135 & 150 & 165 & 180 & 195 & 210 & 225 \\
31 & 46 & 61 & 76 & 91 & 106 & 121 & 136 & 151 & 166 & 181 & 196 & 211 & 226 & 241 & 0
\end{bmatrix}.
$$

It is easy to get the corresponding position pairs through comparing P and C. For example, according to the pixel with gray level 10, it is permuted from [10,0] to [9,11]. According to gray level 127, it is permuted from [15,7] to [5,10].

Thus, in the original chaotic map, the original position (x_k, y_k) and the permuted position (x_{k+1}, y_{k+1}) are both known, and the attacker's aim is computing the parameters a and b. In this case, it is easy to get the following equations.

$$
\begin{cases}
x_{k+1} = x_k + by_k + k_0 M, & k_0 = -M+1,\dots,0,\dots,M-1 \\
y_{k+1} = ax_k + (ab+1)y_k + k_1 M, & k_1 = -M+1,\dots,0,\dots,M-1
\end{cases}.
$$

Here, k_0 and k_1 are determined by the values of a and b. Then, solving these equations, we get

$$\begin{cases} a = (y_{k+1} - y_k - k_1 M) / (x_k - k_0 M), & k_1, k_0 = -M+1,...,0,...,M-1 \\ b = (x_{k+1} - x_k - k_0 M) / y_k, & k_0 = -M+1,...,0,...,M-1 \end{cases}.$$

Thus, the suitable a and b in the range of $[0,M)$ can be obtained. As can be seen, in this attack, only one pixel's position pair is required. And only one pixel is computed iteratively in order to get the suitable a and b. Thus, the attack's computing complexity is greatly decreased.

If the chaotic map is iterated for more times, it is also easy to be broken with the select-plaintext attack. Similarly, some other permutation-only encryption algorithms are also not secure enough. Generally, these permutation operations should be combined with some other operations, such as diffusion [5, 6], in order to improve the system's security.

11.3 Flaws in Perceptual Security

Perceptual security is important to multimedia encryption. Generally, the significant parameters in media content should be encrypted in order to make the content unintelligible. The significant parameters are sensitive to media content's intelligibility. That is, a slight change in the parameters will cause great changes in the media content. In the chapter on partial encryption, some examples of parameter selection are presented. Here, two flaws in parameter selection will be discussed. First, the sensitive or significant parameter does not denote the parameter with large volumes. Second, the perceptual security should be measured by objective metrics.

11.3.1 Level Encryption in Run-Level Coding

The significant parameter may be the short parameter but not the long one. Taking the run-length coding used in a PCX image [10], for example, each line of the image is encoded with run-length coding that produces the (Level,Run) pairs. Here, Level denotes the pixel's gray level, and Run denotes the number of adjacent pixels that have the same Level. Taking an 8-bit image for example, the Level is composed of 8 bits, while the Run is composed of 5 bits. That is, the image is encoded into the sequence of (Level,Run) pairs, that is, (x_0,p_0), (x_1,p_1), ..., (x_{n-1},p_{n-1}). Here, $0 \le x_i < 256$, $0 \le p_i < 64$ ($i = 0, 1, ..., n-1$).

As can be seen, in the two parameters, the first one (Level) is longer than the second one (Run). However, it does not mean that the first one is preferred to be

Original

Encrypted

Figure 11.1   The images encrypted by Level encryption.

encrypted over the second one. For example, only the Level in each pair is encrypted according to

$$x_i' = \mathrm{E}(x_i, K),$$

where $\mathrm{E}()$ is the encryption algorithm and K is the key. Figure 11.1 shows the images encrypted by Level encryption. Here, the AES cipher is used as the encryption algorithm, and the same key is used for the whole image. As can be seen, the encrypted image's border information is still intelligible. In comparison, encrypting the Run obtains high perceptual security, which has been proved in the chapter on partial encryption.

11.3.2   Statistical Model-Based Attack on Sign Encryption

Considering that multimedia encryption emphasizes content encryption, some attacks aim to recover the cipher image's intelligibility under the condition of knowing only cipher images. Statistical model-based attack [11] is an example of this kind of attack. In this attack, the degradation of the cipher image is reduced by making use of a statistical model of the image. Generally, it is realized according to the following steps.

- First, the statistical model of the initial cipher image is constructed. Generally, the unencrypted part of the image is used to tell the estimation information.
- Second, the initial cipher image is recovered by changing some of the encrypted part. This operation is similar to the brute-force attack, but only one pixel or bit is changed in each step.
- Third, a metric is defined to measure the difference between the constructed version and a recovered version.
- Fourth, if the difference is smaller than T (a predetermined threshold), then the recovered cipher image is regarded as the initial cipher image, the second, third and fourth steps are followed, and they are repeated until the difference is small enough. Otherwise, the second, third and fourth steps are repeated directly.

Taking discrete cosine transform (DCT) sign encryption for example, the image is partitioned into blocks, each block is transformed by DCT, and the DCT block's AC coefficients are encrypted by sign encryption. In the statistical model-based attack, the amplitudes of the DC and AC coefficients are used to construct a statistical optimal image, and the image is recovered by changing the AC coefficients' signs. Figure 11.2 shows the images recovered by this attack, where 70% of the signs are recovered. As can be seen, the image's quality can be greatly improved.

Note that, in this attack, the key step is to construct an image with the cipher image's statistical model. The attack works well when the cipher image is degraded to some extent. But, if the cipher image is degraded greatly, the attack does not work. Thus, to resist this attack, the encrypted image should have low quality, such as a small peak signal-to-noise ratio.

Original   Encrypted   Recovered

Figure 11.2   The images recovered by statistical model-based attacks.

11.4 Replacement Attack

In partial encryption, some parameters are encrypted, while the others are left unencrypted. Intuitively, some attackers try to recover the plain media by replacing the encrypted parameters with some others [12]. This kind of attack is named replacement attack. Suppose the original media content P is composed of two parameters, X and Y. Among them, X is encrypted into Z, while Y is left unchanged. Thus, the cipher-media C is generated according to

$$\begin{cases} P = X \| Y \\ Z = E(X, K), \\ C = Z \| Y \end{cases}$$

where E() is the encryption algorithm and K is the key. The replacement attack means to generate the new cipher-media C' under the condition of knowing only Z and Y, which is described by

$$C' = X' \| Y = G(Z, Y) \| Y .$$

Here, $G(Z, Y)$ is the method to generate the new X with the help of Z or Y. The new X, named X', replaces the original X.

According to the method G(), replacement attacks can be classified into two types, direct replacement and correlation-based replacement.

- Direct replacement is used to break multimedia encryption algorithms, which means to replace some of the encrypted data with other ones in order to reconstruct the plain media content under the condition of knowing only the cipher media content. In this case, $X' = G(Z)$. That is, the unencrypted parameter Y is not used.
- Correlation-based replacement is similar to direct replacement. The difference is that some of the encrypted data is replaced by unencrypted data. In this case, $X' = G(Y)$. That is, the unencrypted parameter Y is used to generate X'.

11.4.1 Replacement Attack on Bit-Plane Encryption

In bit-plane encryption [12], the image is partitioned into bit-planes, and the most significant bit-planes are encrypted, while the others are left unencrypted. With respect to this encryption scheme, two kinds of replacement attacks are described, as follows.

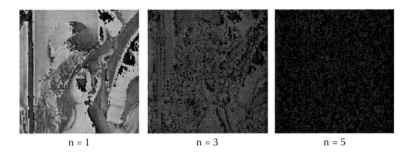

n = 1   n = 3   n = 5

Figure 11.3 The images recovered by direct bit-plane replacement.

In direct replacement, the most significant bit-planes are replaced by zeros. Taking an 8-bit image for example, the bit-planes numbered from 0 to 7 are ordered from the most significant one to the least significant one. By replacing the first n ($n = 1, 2, 3, 4, 5$) bit-planes with zeros, images with different quality can be obtained, as shown in Figure 11.3.

In correlation-based replacement, the most significant bit-planes are predicted by referencing other ones. For example, if the bit in the unencrypted bit-plane is "1," then the corresponding one in the encrypted bit-plane is estimated as "1," otherwise, as "0." Figure 11.4 shows the replaced result when $n = 3$ and $n = 5$. As can be seen, the correlation-based method results in higher quality than the direct method.

11.4.2 Replacement Attack on Frequency Band Encryption

In frequency band encryption, such as band encryption in the wavelet domain, the media content may also be recovered by replacement attack. For example, in

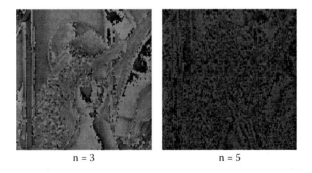

n = 3   n = 5

Figure 11.4 The images recovered by correlation-based bit-plane replacement.

Original   Encrypted   Replaced

Figure 11.5   The images recovered by frequency band replacement.

five-level wavelet transformation, if only the lowest frequency band is encrypted, the image produced, shown in Figure 11.5(center) can be recovered by replacement attack (right). Here, the coefficients in the lowest frequency band are all replacing the certain value 20.

11.4.3   Replacement Attack on Intra-Prediction Mode Encryption

The MEPG4 AVC/H.264 video encryption scheme is proposed in [13], which encrypts only the intra-prediction modes (IPMs) during video compression. By replacing the scrambled IPMs with a certain value [14], the video quality can be improved. Figure 11.6 shows the original video, encrypted video and recovered video. Here, the IPMs are all replaced by the fourth mode.

Original   Encrypted   Replaced

Figure 11.6   The videos recovered by IPM replacement.

11.5 Summary

In this chapter, the general security requirements of multimedia encryption are presented, and some typical attacks on existing encryption algorithms are presented. Note that multimedia data is more complicated than we imagine, and thus, multimedia encryption may face more potential attacks, which need to be resisted by continuous research work.

References

[1] J. A. Buchmann. 2001. *Introduction to Cryptography.* New York: Springer-Verlag.

[2] National Institute of Standards and Testing (NIST). *Recommendation for the Triple Data Encryption Algorithm (TDEA) Block Cipher.* Special Publication 800-67. Gaithersburg, MD: NIST.

[3] FIPS 197. 2001 (November). Advanced Encryption Standard (AES).

[4] M. S. Anoop. 2007. *Public Key Cryptography: Applications, Algorithms and Mathematical Explanations.* India: Tata Elxsi.

[5] S. Lian, J. Sun, and Z. Wang. 2005. Security analysis of a chaos-based image encryption algorithm. *Physica A: Statistical and Theoretical Physics* 351(2-4): 645–661.

[6] S. Lian, J. Sun, and Z. Wang. 2005. A block cipher based on a suitable use of the chaotic standard map. *International Journal of Chaos, Solitons and Fractals* 26(1): 117–129.

[7] S. Lian. 2007. Multimedia content protection based on chaotic neural networks. In *Computational Intelligence in Information Assurance and Security,* ed. Nadia Nedjah, Ajith Abraham, and Luiza de Macedo Mourelle. Berlin: Springer.

[8] D. Karras, and V. Zorkadis. 2003. On neural network techniques in the secure management of communication systems through improving and quality assessing pseudorandom stream generators. *Neural Networks* 16(5-6): 899–905.

[9] F. Pichler, and J. Scharinger. 1996. Finite Dimensional Generalized Baker Dynamical Systems for Cryptographic Applications. Lecture Notes in Computer Science, 1030, 465–476.

[10] ZSoft PCX File Format Technical Reference Manual. http://www.qzx.com/pc-gpe/pcx.txt

[11] A. Said. 2005. Measuring the strength of partial encryption schemes. In *Proceedings 2005 IEEE International Conference on Image Processing (ICIP 2005),* September 11–14, Vol. 2, 1126–1129.

[12] M. Podesser, H. P. Schmidt, and A. Uhl. 2002. Selective bitplane encryption for secure transmission of image data in mobile environments. In *Proceedings 5th IEEE Nordic Signal Processing Symposium (NORSIG 2002)* [CD-ROM], Tromso-Trondheim, Norway, October.

[13] J. Ahn, H. Shim, B. Jeon, and I. Choi. 2004. Digital Video Scrambling Method Using Intra Prediction Mode. Lecture Notes in Computer Science, 3333, 386–393.

[14] S. Lian, J. Sun, G. Liu, and Z. Wang. 2007. Efficient Video Encryption Scheme Based on Advanced Video Coding. Multimedia Tools and Applications. Preprint.

Chapter 12

Some Principles for Secure Multimedia Encryption

12.1 Shannon's Information Theory of Secrecy System

Shannon [1] defined a secrecy system as a communication process, and used information entropy theory to model the process. As shown in Figure 12.1, the cipher's encryption process is defined as

$$C = \mathrm{E}(P, K),\qquad (12.1)$$

where P, $\mathrm{E}()$, K, and C are the plaintext, the cipher's encryption algorithm, key, and ciphertext, respectively.

In Shannon's theory [1], the cipher's plaintext P and ciphertext C are regarded as the communication channel's input and output, respectively. Then, the entropy of the input is defined as

$$H(P) = -\sum_{i=0}^{n-1} p_i \log p_i,$$

where p_i ($i = 0, 1, \ldots, n-1$) is P's distribution probability. Entropy is really a notion of self-information, that is, the information provided by a random process about itself. Conditional entropy is defined as

$$H(P \mid C) = -\sum_{i=0}^{n-1} p_i \log p_{P\mid C, i},$$

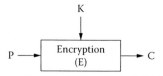

Figure 12.1 The cipher's encryption process.

where $p_{P|C,j}$ denotes P's conditional probability with respect to C. This information can be used to measure the noise that changes P to C. Average mutual information is defined as

$$I(P,C) = H(P) - H(P|C) = H(C) - H(C|P). \tag{12.2}$$

Mutual information is a measure of the information contained in one process about another process. It is in close relation with the correlation between the two variables. Additionally, it can be defined as

$$I(P,C) = H(P) + H(C) - H(P,C), \tag{12.3}$$

where $H(P,C)$ is the pair information of P and C.

Based on definitions, Shannon gave some conditions that a general encryption process should satisfy.

First, it is easy to compute C from both P and K, and K is selected independently from P. Thus, we get

$$\begin{cases} H(C|P,K) = 0 \\ H(K|P) = H(K) \end{cases}. \tag{12.4}$$

Second, the difficulty of ciphertext-only attack on a cryptosystem is measured by $H(P,K|C)$, and the difficulty of known-plaintext attack on a cryptosystem is measured by $H(K|P,C)$.

Third, a cryptosystem has ideal security, if it meets the following condition.

$$\begin{cases} H(P,K|C) = H(K|C) + H(P|C) = H(K) + H(P) \\ H(K|P,C) = H(K) \end{cases}. \tag{12.5}$$

In the following discussion, the entropy-based theory will be used to analyze the three kinds of multimedia encryption schemes and present some principles for designing a secure scheme.

12.2 Principles for Secure Complete Encryption

12.2.1 General Models of Complete Encryption

As mentioned in Chapter 4, complete encryption is classified into two types, encryption before compression, and encryption after compression. The former, as shown in Figure 12.2(a), is defined as

$$C = E(P,K),$$

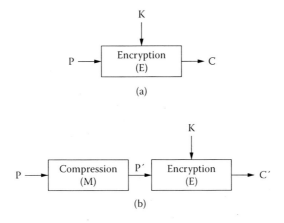

Figure 12.2   General models of complete encryption.

where P, E(), K and C are the original media, the encryption operation, key and cipher-media, respectively. The latter, as shown in Figure 12.2(b), is defined as

$$C' = E(P', K) = E(M(P), K),$$ (12.6)

where P' and M() are the compressed media and compression operation, respectively.

12.2.2 Encryption before or after Compression

The two models may get different ideal security. For the first model, it is easy to get the ideal security.

$$H(P, K \mid C) = H(K) + H(P).$$ (12.7)

For the second model, we get the ideal security.

$$H(P, K \mid C') = H(P', K \mid C') = H(K) + H(P').$$ (12.8)

Generally, for the compression process, the following condition is satisfied.

$$H(P') - H(P) \geq 0.$$ (12.9)

This is because there is much more redundancy in the raw media than the compressed data. Thus, we get

$$H(P, K \mid C') \geq H(P, K \mid C).$$ (12.10)

(a) Original	(b) Encrypt before compression	(c) Encrypt after compression

Figure 12.3   Images encrypted by different complete encryption methods.

According to this point, we get the following principle.

> Principle 12.1. In complete encryption, encrypting the compressed media data will obtain higher security against ciphertext-only attacks than encrypting the raw data.

This principle can also be shown by the simple experiment that encrypts the raw image and JPEG [2] compressed image, respectively, as shown in Figure 12.3. Compared with Figure 12.3(c) obtained by encrypting the compressed image with Advanced Encryption Standard (AES) [3], Figure 12.3(b) obtained by encrypting the raw image with AES, explores some secret information.

12.2.3 *Example of Secure Complete Encryption*

It is easy to encrypt the compressed media data directly. However, the encryption efficiency should be considered especially when real-time applications are required. Taking media streaming, for example, the compressed media data is packaged into packets that are transmitted and decoded in real time. To meet this application, the scheme is proposed to encrypt the packets in a lightweight manner. First, the packet is partitioned into subpackets in a random manner, then the subpackets are encrypted with Video Encryption Algorithm (VEA) [4], as shown in Figure 12.4.

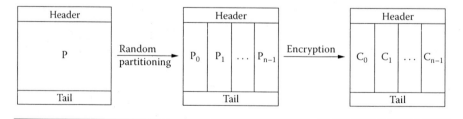

Figure 12.4   Packet encryption based on Video Encryption Algorithm (VEA).

12.2.3.1 Random Packet Partitioning

Set $P = p_0 p_1 \ldots p_{N-1}$ (p_i, $i = 0, 1, \ldots, N-1$, is the bit in the packet) the original media data in a packet, P_0, P_1, $\ldots$ and P_{n-1} the n subpackets of the packet P. To partition the packet into subpackets, a random sequence $R = r_0 r_1 \ldots r_{N-1}$ ($0 \leq r < n$) is generated under the control of K. Then, in the partitioning, which subpacket the bit p_j ($j = 0, 1, \ldots, N-1$) belongs to is decided by the following method.

$$p_j \begin{cases} \in P_i, & p_j = i \, (j = 0, 1, \ldots, N-1, i = 0, 1, \ldots, n-1) \\ \notin P_i, & p_j \neq i \, (j = 0, 1, \ldots, N-1, i = 0, 1, \ldots, n-1) \end{cases}. \tag{12.11}$$

12.2.3.2 Packet Encryption

For the n subpackets, only the first one is encrypted by a traditional strong cipher, such as 3DES or AES [5], and the other $n-1$ subpackets are encrypted by bitwise exclusive or (XOR) operation. The encryption operations shown in Figure 12.5(a) are defined as

$$\begin{cases} C_0 = E(P_0, K), \\ C_i = P_i \oplus P_{i-1}, & (i = 1, 2, \ldots, n-1) \end{cases}. \tag{12.12}$$

Here, E() is the encryption operation of the adopted cipher, $\oplus$ is the XOR operation, and C_0, C_1, $\ldots$ and C_{n-1} are the encrypted subpackets corresponding to P_0, P_1, $\ldots$ and P_{n-1}, respectively.

12.2.3.3 Decryption

The partial decryption process is symmetric to the encryption process. First, the packet is partitioned into n subpackets. Then, for the n subpackets, only the first one is decrypted by a traditional strong cipher, and the other $n-1$ subpackets are decrypted by XOR operation. The decryption operations shown in Figure 12.5(b) are defined as

$$\begin{cases} P_0 = D(C_0, K), \\ P_i = C_i \oplus P_{i-1}, & (i = 1, 2, \ldots, n-1) \end{cases} \tag{12.13}$$

Here, the parameters are similar to the ones in encryption operations.

12.2.3.4 Encryption Efficiency

For each packet, there are n subpackets, and each subpacket is composed of $64t$ bits ($t > 0$ and t is an integer). In all the subpackets, only the first one is encrypted by

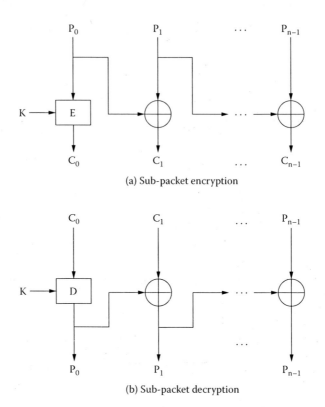

(a) Sub-packet encryption

(b) Sub-packet decryption

Figure 12.5 Subpacket encryption and decryption operations.

a traditional cipher, while the others are encrypted by XOR operation. It is tested that, for each 64-bit plaintext, the encryption time ratio between XOR and AES is about 0.06. Thus, for the whole packet, the time ratio (TR) between VEA encryption and complete AES encryption is

$$\text{TR} = \frac{tT + 0.06(n-1)tT}{ntT} = \frac{1}{n} + 0.06\frac{n-1}{n}. \tag{12.14}$$

As shown in Figure 12.6, the time ratio decreases with increase in n. The ratio is smaller than 0.5 when n is no smaller than 3.

12.3 Principles for Secure Partial Encryption

12.3.1 Model of Partial Encryption

The proposed partial encryption model is shown in Figure 12.7, which combines the encryption process with the compression process, and is composed of media compression, parameter encryption, and media decompression. Here, P, K, and C

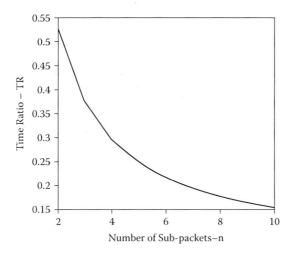

Figure 12.6 Time ratio between Video Encryption Algorithm (VEA) and complete Advanced Encryption Standard (AES) encryption.

are the plain-media, key and cipher-media, respectively. X and Y are the parameters in the data stream, among which X is encrypted into Z, while Y is left unchanged. The encryption process is defined as

$$E(X,K) = Z, \qquad (12.15)$$

where $E()$ is the encryption operation. Without losing the generality, the data stream composed of two parameters is investigated. If the data steam is composed of more parameters, similar results can be obtained.

12.3.2 Principles for Secure Encryption

12.3.2.1 Cryptographic Security

In parameter encryption, the cipher $E()$ is used to encrypt the parameter X, whose security is in close relation with the system's security.

According to Section 12.1, the security against known-plaintext attack is $H(K|P,C)$. Thus, in the proposed partial encryption scheme, we get

$$H(K \mid P,C) = H(K \mid X,Y,Z). \qquad (12.16)$$

According to Equation (12.4), we get

$$H(K \mid P) = H(K \mid X,Y) = H(K). \qquad (12.17)$$

According to Equations (12.16) and (12.17), we get

$$H(K \mid P,C) = H(K \mid Z). \tag{12.18}$$

Thus, we get the following principle.

> Principle 12.2. In the partial encryption scheme, the scheme's security against known-plaintext attack depends on the security of the adopted cipher.

Thus, to design a secure partial encryption scheme, the cipher with high security should be selected.

12.3.2.2 Parameter Independence

In the partial encryption scheme, the relation between the encrypted parameter X and the unencrypted parameter Y affects the scheme's security.

According to Section 12.1, an encryption algorithm's security against ciphertext-only attack is $H(K,P|C)$. Thus, in the partial encryption scheme, the following equation holds:

$$H(K,P \mid C) = H(K,X,Y \mid Z,Y) = H(K,X \mid Z,Y). \tag{12.19}$$

According to Shannon's theory, we get

$$H(K,X \mid Z,Y) = H(K,X \mid Z) - I(K,X,Y \mid Z). \tag{12.20}$$

Here, $I(A,B|C)$ means the mutual information between A and B under the condition of C. According to Equation (12.4), K is independent of X and Y, so we get

$$I(K,X,Y \mid Z) = I(X,Y \mid Z). \tag{12.21}$$

Thus, from Equations (12.19), (12.20) and (12.21), we get

$$H(K,P \mid C) = H(K,X \mid Z) - I(X,Y \mid Z) \le H(K,X \mid Z). \tag{12.22}$$

Here, $H(K,P|C)$ gets the maximal value $H(K,X|Z)$ if and only if $I(X,Y|Z) = 0$, that is, X is independent from Y. Thus, we get the following principle.

> Principle 12.3. In the partial encryption scheme, its security against ciphertext-only attack gets the maximal value, if and only if the encrypted parameter is independent of the unencrypted parameter.

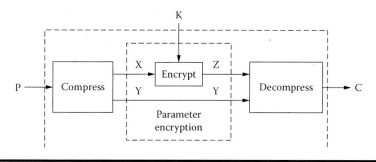

Figure 12.7   A model of partial encryption.

Therefore, to design a secure partial encryption scheme, the encrypted parameter should be independent of the unencrypted parameter. This means resists attacks that make use of the correlation between various parameters, such as the correlation-based replacement attack [6].

12.3.2.3 Perceptual Security

Perceptual security denotes the encrypted media content's intelligibility, which is in close relation with the sensitivity of the encrypted parameters.

For the partial encryption scheme shown in Figure 12.7, we define parameter-sensitivity of the encoding process as

$$S_e = \frac{E_n(P + \Delta P) - E_n(P)}{\Delta P} = \frac{\Delta X}{\Delta P}. \tag{12.23}$$

Here, $E_n()$ denotes the compression process, and only parameter X is considered when Y is left unencrypted. ΔP and ΔX are the slight disturbance on P and X, respectively. Similarly, the parameter-sensitivity of the decompression process is defined as

$$S_d = \frac{D_e(Z + \Delta Z) - D_e(Z)}{\Delta Z} = \frac{\Delta C}{\Delta Z}. \tag{12.24}$$

Here, $D_e()$ denotes the decoding process, and only parameter Z is considered when Y is left unencrypted. ΔZ and ΔC are the slight disturbance on Z and C, respectively. In fact, for a certain codec, $P = C$ and $X = Z$ are satisfied, so we can get the following condition from Equations (12.23) and (12.24).

$$S_e \times S_d = 1. \tag{12.25}$$

For the adopted cipher, its plaintext-sensitivity is defined as

$$S_p = \frac{E(X + \Delta X, K) - E(X, K)}{\Delta X} = \frac{\Delta Z}{\Delta X}. \tag{12.26}$$

Here, $E(A, B)$ denotes the encryption process in which A is the plaintext and B is the key. Similarly, its key-sensitivity is defined as

$$S_k = \frac{E(X, K + \Delta K) - E(X, K)}{\Delta K} = \frac{\Delta Z}{\Delta K}. \tag{12.27}$$

Here, ΔK is the slight disturbance on K. Therefore, we get the plaintext-sensitivity of the partial encryption scheme.

$$S_{pp} = \frac{\Delta C}{\Delta P} = \frac{\Delta Z \times S_d}{\Delta P} = \frac{\Delta X \times S_p \times S_d}{\Delta P}$$

$$= \frac{\Delta P \times S_e \times S_p \times S_d}{\Delta P} = S_e \times S_p \times S_d = S_p. \tag{12.28}$$

Similarly, the key-sensitivity of the partial encryption scheme is

$$S_{kp} = \frac{\Delta C}{\Delta K} = \frac{\Delta Z \times S_d}{\Delta K} = \frac{\Delta K \times S_k \times S_d}{\Delta K} = S_k \times S_d. \tag{12.29}$$

In the partial encryption scheme, set the parameter-sensitivity of compression and decompression processes as S_e and S_d, respectively, and the key-sensitivity and plaintext-sensitivity of the adopted cipher as S_k and S_p, respectively, then the plaintext-sensitivity of the partial encryption scheme is $S_{pp} = S_p$, and its key-sensitivity is $S_{kp} = S_k \times S_d$. As can be seen, the plaintext-sensitivity of the partial encryption scheme is determined by the plaintext-sensitivity of the adopted cipher. The key-sensitivity of the partial encryption scheme is determined by the parameter-sensitivity of the decompression process. Thus, we get the following principle.

> Principle 12.4. Because the key-sensitivity increases with the rise of the parameter-sensitivity, the parameters with high sensitivity should be encrypted in order to resist sensitivity-based attacks, such as direct replacement [6] and statistical model-based quality improvement [7].

12.3.2.4 Summary of the Principles and Means

According to the above analyses, the cryptographic security and perceptual security should be considered when designing a secure multimedia encryption scheme. Additionally, for partial encryption, the parameter security needs to be considered. To confirm the security requirements, some means and principles are proposed, as listed in Table 12.1. First, the strong cipher should be adopted to keep high

Table 12.1   Security Requirements and Means of Multimedia Encryption

Security Requirement	Means to Confirm the Security
Cryptographic security	Use strong ciphers
Perceptual security	Encrypt the significant parameters
Parameter security in partial encryption	Keep the encrypted parameters independent from the unencrypted ones

cryptographic security. Second, the significant parameters should be encrypted in order to keep perceptual security. Additionally, in the case of partial encryption, the encrypted parameters should be independent from the unencrypted ones.

12.3.3   Example Based on JPEG2000 Image Encryption

In JPEG2000 codec, the data stream is composed of three parameters, subband, bit-plane, and coding pass. The image is transformed into different frequency bands that represent different fidelity or resolution. Additionally, each subband is partitioned into a number of code blocks, and each code block is encoded bit-plane by bit-plane from the most significant one to the least significant one. In addition, each bit-plane is encoded with three passes, among which, the significant pass encodes some significant coefficients' signs, the refinement pass encodes some coefficients' bit value, and the cleanup pass encodes both some significant coefficients' signs and coefficients' position information. Below, we test the sensitivity of the proposed three parameters, that is, subband, bit-plane or code pass, respectively, select the parameters according to the independence, select a suitable strong cipher, and construct the secure partial encryption scheme.

12.3.3.1   Parameter Sensitivity

In L-level wavelet transform, an image is transformed into $3L + 1$ subbands. For example, the five-level wavelet transform produces the following subbands: LL_5, LH_5, HL_5, HH_5, …, LH_1, HL_1 and HH_1. Each subband has different sensitivity to the image's understandability. We encrypt each of the subbands, and compute the peak signal-to-noise ratio of the corresponding image. The experimental results are shown in Table 12.2 (only the lowest five subbands are listed). As can be seen, the image's quality is the worst when the lowest frequency (LL_5) is encrypted. However, for other frequency bands, the image's quality is similar to each other. It tells that the subband of the lowest frequency has higher sensitivity to the images' understandability, and it is favored to be encrypted.

Similarly, each bit-plane of the code block has different sensitivity to the image's understandability. Taking the same image for example, we encrypt each of the bit-planes and get the encrypted image. The sensitivity is shown in Table 12.3. Thus, the

Table 12.2 Subband Sensitivity

				Image Peak Signal-to-Noise Ratio (dB)				
Image	Size	Wavelet Level	Subband-1	Subband-2	Subband-3	Subband-4	Subband-5
Lena	128 × 128	4	10.6	19.4	18.7	20.1	18.8
Couple	256 × 256	5	8.7	19.8	17.9	19.3	19.2
Cameraman	256 × 256	5	9.3	21.2	20.4	21.5	20.1
Goldhill	512 × 512	6	7.6	20.3	19.6	20.3	19.5
Boats	512 × 512	6	7.3	19.5	19.3	19.2	20.2

Table 12.3 Bit-Plane Sensitivity

Image	Size	Wavelet Level	Image Peak Signal-to-Noise Ratio (dB)					
			Bitplane-0	Bitplane-3	Bitplane-6	Bitplane-8	Bitplane-10	
Lena	128 × 128	4	44.8	34.5	24.3	22.1	19.5
Couple	256 × 256	5	43.9	33.2	23.1	20.3	17.9
Cameraman	256 × 256	5	45.2	35.7	24.8	21.9	19.2
Goldhill	512 × 512	6	44.6	33.6	24.4	22.3	19.3
Boats	512 × 512	6	44.3	34.5	23.9	20.8	18.2

significant bit-planes (from the sixth to the tenth) are more sensitive to the images' understandability than less significant ones, which are preferred to be encrypted.

In bit-plane encoding, the significant pass, refinement pass and cleanup pass have different sensitivity to the images' understandability, which is tested with the similar method, as shown in Table 12.4. As can be seen, the curve corresponding to the cleanup pass is the lowest one. Thus, the cleanup pass is more sensitive than other two passes, and it is preferred to be encrypted.

12.3.3.2 Parameter Independence

In the proposed three aspects (subband, bit-plane, and encoding-passes), bit-planes are independent from each other, encoding-passes are also independent from each other, but subbands are not. In wavelet transform, the subbands in different layers are dependent on each other. That is, the subbands in the lower layer can be recovered from the ones in the higher layer. Thus, it is not secure to leave all the subbands in a layer unencrypted. For example, in a five-level wavelet transform, the subbands LH_1, HL_1 and HH_1 should not be left unencrypted because they can help to recover the subbands LH_2, HL_2 and HH_2. In each code block, the bit-planes are independent from each other, so they can be encrypted selectively. For each bit-plane, the three encoding-passes are also independent from each other, which makes them able to be selectively encrypted.

12.3.3.3 Selection of the Cipher

In JPEG2000 encoding, the encoding-passes are often of variable size, the block ciphers in CTR mode or stream ciphers are preferred, for example, AES CTR [8] or RC4 [9].

12.3.3.4 The Partial Encryption Scheme

According to the above analyses, we encrypt a JPEG2000 image with the following method. First, the transformed frequency bands are partitioned into three parts [10], low-frequency, middle-frequency, and high-frequency. Taking a three-level wavelet transform for example, the low-frequency part is LL_3, the middle-frequency part includes LH_3, HL_3, HH_3, LH_2, HL_2 and HH_2, and the high-frequency part includes H_1, HL_1 and HH_1. For higher-level (e.g, four- or five-level) wavelet transform, similar frequency parts can be produced. Then, for each part, the media data is encrypted as follows.

- Low-frequency part—All the subbands in the low-frequency part are encrypted. For each code block in the subbands, the four most significant bit-planes are encrypted. And for each bit-plane, the three encoding-passes are all encrypted.

Table 12.4 Encoding-Pass Sensitivity

Image	Size	Wavelet Level	Image Peak Signal-to-Noise Ratio (dB)		
			Significance Pass	Refinement Pass	Cleanup Pass
Lena	128 × 128	4	10.1	17.4	12.7
Couple	256 × 256	5	10.7	18.2	12.9
Cameraman	256 × 256	5	9.8	16.2	12.4
Goldhill	512 × 512	6	11.6	20.3	13.6
Boats	512 × 512	6	9.3	15.9	12.2

- Middle-frequency part—All the subbands in the middle-frequency part are encrypted. For each code block in the subbands, the six most significant bit-planes are encrypted. And for each bit-plane, only the cleanup pass is encrypted.
- High-frequency part—All the subbands in the high-frequency part are encrypted. For each of them, the three most significant bit-planes are encrypted. And for each bit-plane, only the cleanup pass is encrypted.

12.3.3.5 Performance Evaluation

In the discussion below, the proposed encryption scheme's security against such attacks as correlation-based replacement attack, direct replacement attack and statistical model-based attack is analyzed and compared with some existing schemes. Here, the block cipher of AES CTR is used, which confirms the cryptographic security. Additionally, the efficiency of the encryption scheme in terms of time is tested.

12.3.3.5.1 Perceptual Security

In the proposed scheme, the significant parameters are encrypted, which make the encrypted image unintelligible. Taking Lena (256×256, gray, five-level wavelet transform), for example, the encryption result is shown in Figure 12.8. It shows that the proposed encryption scheme provides high perceptual security.

12.3.3.5.2 Security against Replacement Attack

Replacement attack [6] includes two types, direct replacement attack and correlation-based replacement attack. Direct replacement replaces the encrypted parameters with some other parameters directly. The scheme's security against direct replacement is determined by the sensitivity of the encrypted parameters. According

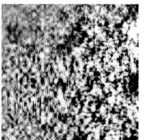

Original   Encrypted

Figure 12.8   Result of partial image encryption.

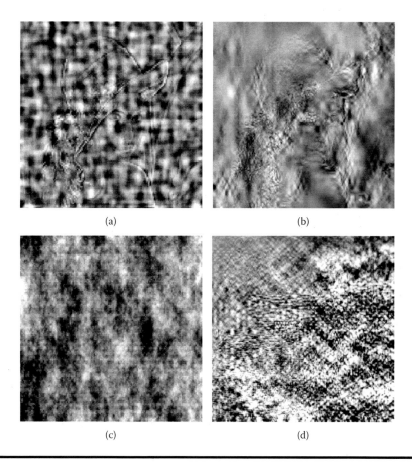

Figure 12.9 Images recovered by replacement attack. (a) The image recovered by direct replacement in the existing encryption scheme. (b) The image recovered by correlation-based replacement in the existing encryption scheme. (c) The image recovered by direct replacement in the proposed scheme. (d) The image recovered by correlation-based replacement in the proposed scheme.

to the second principle, if the parameters with high sensitivity are all encrypted, then the scheme's security can be confirmed. Correlation-based replacement attack makes use of the relation between the encrypted parameter and the unencrypted parameter to recover the encrypted one and thus to make the encrypted image intelligible. According to the third principle, if the encrypted parameter bears little relation to the unencrypted parameter, the scheme's security can be confirmed. Here, we compare the proposed scheme with the existing encryption scheme [11, 12] that encrypts only the low-frequency and middle-frequency parts while leaving the high-frequency part unchanged. Taking Lena for example, the images recovered by

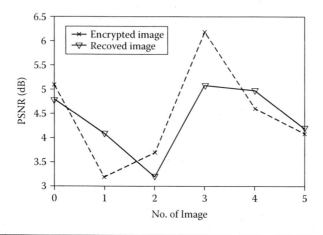

Figure 12.10 Quality improvement by statistical model-based attack.

replacement attacks are shown in Figure 12.9. As can be seen, the recovered images in the proposed scheme have lower quality compared with the ones in the existing scheme. Thus, the proposed scheme is more secure against replacement attack than the existing scheme.

12.3.3.5.3 Security against Statistical Model-Based Attack

In the proposed scheme, the parameters with high sensitivity are encrypted, and thus, the encrypted image is degraded greatly. According to the properties of statistical model-based attack [7], the attack is practical when the encrypted image is not greatly degraded. Thus, with respect to the proposed encryption scheme, the statistical model-based attack does not work. Using this attack to recover the images encrypted by the proposed scheme, the results are shown in Figure 12.10. Here, various images are tested, including Airplane, Cameraman, Lena, Bridge, Couple, and Baboon. They are ordered as: 0-Airplane, 1-Cameraman, 2-Lena, 3-Bridge, 4-Couple, and 5-Baboon. As can be seen, little quality improvement is obtained.

12.3.3.5.4 Time Efficiency

The proposed encryption scheme encrypts only some sensitive data. The encryption data ratio and encryption time ratio are tested, as shown in Table 12.5. The former denotes the ratio between the encrypted data and the whole data stream, while the latter denotes the ratio between the encryption time and encoding time. As can be seen, the encryption data ratio is often no more than 15%, and the encryption

Table 12.5 Time Efficiency of the Partial Encryption Scheme

Image	Size	Colorful/Gray	Encryption Data Ratio	Encryption Time Ratio
Lena	128 × 128	Gray	10.5%	7.7%
Boat	256 × 256	Gray	8.9%	9.2%
Cameraman	256 × 256	Gray	11.4%	8.0%
Village	512 × 512	Gray	13.3%	10.3%
Lena	128 × 128	Color	12.1%	7.9%
Jet	256 × 256	Color	12.7%	8.8%
Peppers	256 × 256	Color	11.2%	9.5%
Baboon	512 × 512	Color	13.6%	9.9%

time ratio is often smaller than 15%. Thus, the partial encryption scheme is time efficient and suitable for some real-time applications.

12.4 Principles for Secure Compression-Combined Encryption

12.4.1 Model of Compression-Combined Encryption Scheme

The compression-combined encryption scheme is shown in Figure 12.11(b), which is compared with the normal compression process shown in Figure 12.11(a). According to Section 12.1, the encryption scheme's security against ciphertext-only attack and known-plaintext attack are defined as

$$\begin{cases} H(KP \mid P') = H(KP \mid C) \\ H(K \mid PP') = H(K \mid C) \end{cases}. \tag{12.30}$$

The relation between compression ratio and security is investigated below.

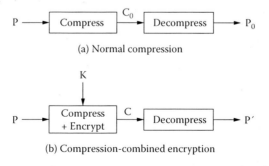

(a) Normal compression

(b) Compression-combined encryption

Figure 12.11 Compression-combined encryption and normal compression.

12.4.2 Relation between Compression Ratio and Security

According to Equation (12.30), the security against known-plaintext attack is

$$H(K \mid C) = H(K) + H(P) - H(C) = H(K) - (H(C) - H(P)). \quad (12.31)$$

Set N_e the ciphertext space and D_e the compression ratio. Then, according to the compression operation, we get

$$D_e = \log N_e - H(P) \geq H(C) - H(P). \quad (12.32)$$

Considering that P is compressed into C, we get $D_e \geq 0$. Thus, according to Equation (12.32), we get

$$\log N_e - H(P) \geq 0. \quad (12.33)$$

According to Equations (12.32) and (12.33), Equation (12.31) is deduced to

$$H(K \mid C) = H(K) - (H(C) - H(P))$$
$$\geq H(K) - (\log N_e - H(P)) = H(K) - D_e. \quad (12.34)$$

When C is in uniform distribution, Equation (12.35) holds:

$$H(C) = \log N_e. \quad (12.35)$$

In this case, the security against known-plaintext attack is

$$H(K \mid C) = H(K) - D_e. \quad (12.36)$$

Thus, the security $H(K|C)$ decreases with increase in D_e. Thus, we get the following principle.

> Principle 12.5. In compression-combined encryption, the compression ratio may contradict the security against known-plaintext attack. A suitable trade-off between them should be obtained to design a good compression-combined encryption scheme.

12.4.3 Equivalent Compression–Encryption Independent Scheme

For some compression-combined encryption schemes, there are equivalent compression-encryption independent schemes. As shown in Figure 12.12, the security

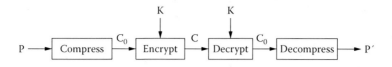

Figure 12.12   The compression-encryption independent scheme.

of the compression-encryption independent scheme against ciphertext-only attack and known-plaintext attack is defined as

$$\begin{cases} H(KP \,|\, C) = H(KC_0 \,|\, C) \\ H(K \,|\, PC) = H(K \,|\, CC_0) \end{cases}.$$
$$(12.37)$$

According to Equation (12.4), we get

$$H(K \,|\, C_0) = H(K) \text{ and } I(C_0, K) = 0. \qquad (12.38)$$

Thus, Equation (12.37) becomes

$$\begin{cases} H(KP \,|\, C) = H(KC_0 \,|\, C) = H(K \,|\, C) + H(C_0 \,|\, C) \\ H(K \,|\, PC) = H(K \,|\, CC_0) = H(K \,|\, C) \end{cases}. \qquad (12.39)$$

As can be seen from Equation (12.39), the security of the original scheme is determined by the security of the equivalent scheme. Thus, we get the following principle.

> Principle 12.6. If the compression-combined encryption scheme has an equivalent compression-encryption independent scheme, then the security of the original scheme can be measured by the security of the equivalent scheme.

12.4.4 Analysis on Existing Compression–Combined Encryption Schemes

According to the above analyses, two principles should be taken care of when designing or evaluating compression-combined encryption schemes. First, the tradeoff between compression ratio and security should be made. Second, the equivalent scheme can be found to evaluate the original scheme's security. Taking existing compression-combined encryption schemes for example, we analyze their properties according to the principles.

12.4.4.1 Tradeoff between Compression Ratio and Security

As mentioned in Chapter 6, there exist some compression-combined encryption schemes, including Huffman tree permutation, random interval selection in arithmetic coding, random coefficient scanning, index encryption, etc. The typical one is random coefficient scanning. In the DCT or wavelet domain, the more the coefficients are permuted, the higher the security is. However, the more the coefficients are permuted, the higher the changed compression ratio is. In the wavelet domain, during complete permutation, subband permutation and quadtree permutation, subband permutation gets a tradeoff. In the DCT domain, segment permutation is able to get a tradeoff if the segments are partitioned in a suitable manner. For Huffman tree permutation, index encryption and random interval selection in arithmetic coding, it is also necessary to get a tradeoff between security and compression ratio, which are not explained one by one.

12.4.4.2 The Equivalent Compression–Encryption Independent Scheme

According to the definition in Chapter 6, the compression-combined encryption scheme can be regarded as the combination of encryption operations and compression operations. Generally, for most combined schemes, it is easy to find the equivalent independent operations. For example, in the scheme based on Huffman tree permutation, the Huffman coding and tree permutation can be isolated. Thus, the scheme's security depends on permutation operation. Similarly, in the scheme based on index encryption, the scheme's security depends on the cipher (stream cipher or block cipher) used to encrypt the indexes. Additionally, in the scheme based on random scanning, the security depends on the coefficient permutation. According to this point, it is easy to tell whether a compression-combined encryption scheme is secure or not. Of course, if you want to design a new scheme, this principle should be used.

12.5 Summary

In this chapter, some principles for secure multimedia encryption are proposed. These principles are constructed on the investigation of the three kinds of encryption schemes. To prove their soundness, some examples are presented. It is expected to provide guidance to researchers or engineers working in this field.

References

[1] C. Shannon. 1949. Communication theory of secrecy systems. *Bell System Technical Journal* 28: 656–715.
[2] W. B. Pennebaker, and J. L. Mitchell. 1993. *JPEG Still Image Compression Standard*. NY: Van Nostrand Reinhold.
[3] FIPS 197. Advanced Encryption Standard (AES). November 2001.

[4] L. Qao, and K. Nahrstedt. 1997. A new algorithm for MPEG video encryption. In *Proceedings First International Conference on Imaging Science, Systems and Technology (CISST'97)*, Las Vegas, NV, July, 21–29.

[5] J. A. Buchmann. 2001. *Introduction to Cryptography.* NY: Springer-Verlag.

[6] M. Podesser, H. P. Schmidt, and A. Uhl. 2002. Selective bitplane encryption for secure transmission of image data in mobile environments. In *Proceedings 5th IEEE Nordic Signal Processing Symposium (NORSIG 2002)* [CD-ROM], Tromso-Trondheim, Norway, October

[7] A. Said. 2005. Measuring the strength of partial encryption schemes. In *Proceedings 2005 IEEE International Conference on Image Processing (ICIP 2005)*, September 11–14, Vol. 2, 1126–1129.

[8] RFC 3686-Using Advanced Encryption Standard (AES) Counter Mode With IPsec Encapsulating Security Payload (ESP). http://rfc.dotsrc.org/rfc/rfc3686.html

[9] B. Schneier. 1996. Section 17.1 RC4. In *Applied Cryptography*, 2nd ed. NY: John Wiley & Sons.

[10] S. Lian, J. Sun, D. Zhang, and Z. Wang. 2004. A selective image encryption scheme based on JPEG2000 codec. Presented at the 2004 Pacific-Rim Conference on Multimedia (PCM2004). Lecture Notes in Computer Science, 3332, 65–72.

[11] T. Uehara. 2001. Combined Encryption and Source Coding. http://www.uow.edu.au/~tu01/CESC.html

[12] S. Wee, and J. Apostolopoulos. 2003. Secure scalable streaming and secure transcoding with JPEG-2000. Technical Report, HPL-2003-117, HP Laboratories, Palo Alto, CA, June 19.

Chapter 13

Multimedia Encryption in Typical Applications

13.1 Secure Media Player

Multimedia encryption techniques can be integrated into the media player. Thus, the player can not only decompress and render media data, but also decrypt media data. For this kind of player, the input data include the decryption key and the encrypted media data. It permits service providers to make their own players with security functionalities embedded. Compared with a normal player, the secure player has a decryption operation. To maintain the original performance, e.g., real-time playing, the decryption operation should be carefully designed.

- The decryption operation and compression operation should fit each other. In the player, the decryption operation and decompression operation can be ordered in different mode, which depends on the encryption method. For example, if the media data is first compressed then encrypted on the sender side, then in the player, the decryption operation is followed by the decompression operation, as shown in Figure 13.1. If the encryption operation is combined with the compression operation, then, in the player, the decryption operation should be combined with the decompression operation. The third case is that media data is first encrypted then compressed. However, this case seldom happens, and thus, it is not considered here.
- The decryption operation should be efficient enough for real-time operations. Considering that, during playing, media data is first decompressed

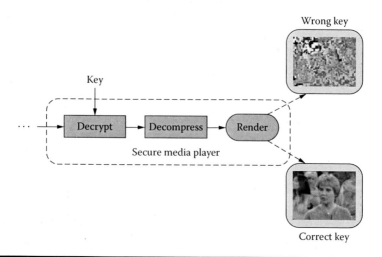

Figure 13.1 Architecture of a secure media player.

and then rendered, the inserted decryption operation may cause delays to the decompression operation. Generally, a symmetric cipher is more widely used, and thus, the encryption operation is symmetric to the decryption operation. Therefore, for the secure player, a media encryption method with high encryption efficiency is preferred.

13.2 Secure Media Streaming

Media streaming is now widely used in real-time entertainment, such as video-on-demand [1], IPTV [2], mobile TV [3], etc. Media streaming has apparent advantages over the download-based techniques. In download-based techniques, a media program should be completely downloaded and stored in the user's device before it can be played back. In contrast, in media streaming, the media program can be played back while it is downloading. According to this property, streaming media aims to provide real-time media content transmission. Secure media streaming provides media streaming services with the security confirmed. As shown in Figure 13.2, media content is first encrypted by the sender and then streamed, and only an authorized user can view the media content. For secure media streaming, two properties are emphasized, security and efficiency.

Security is the basic requirement of secure media streaming. Various methods have been proposed for secure content transmission. However, not all of them are suitable for multimedia data. Taking three typical ones, ISMACryp [4], SRTP [5], and IPSec [6], for example, their properties are introduced here.

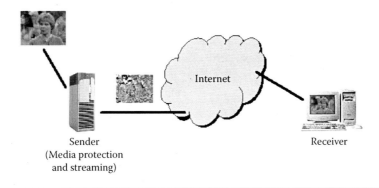

Figure 13.2 Architecture of secure media streaming.

- ISMACryp defines some methods for encrypting and authenticating MPEG-4 [7] data streams. Since MPEG-4 is the codec for video content compression, the ISMACryp works in the application layer.
- SRTP defines some means to encrypt and authenticate RTP (Real-time Transport Protocol) packets. Since RTP is a technique working in the transport layer, the SRTP works in the transport layer.
- IPSec defines some means to encrypt and authenticate IP packets and some methods for secure exchanging. Since IP belongs to the network layer, the IPSec works in the network layer.

Generally, a secure transmission method working in a higher layer can obtain higher security than one working in a lower layer. For example, ISMACryp can realize end-to-end security, while IPSec can only realize peer-to-peer security. Additionally, SRTP and IPSec are designed for general data transmission, but ISMACryp is suitable for multimedia transmission.

Encryption efficiency is a key issue in secure media streaming. Under the architecture designed by ISMACryp, some partial encryption algorithms can be introduced to encrypt the parameters in MPEG4 data stream selectively. Additionally, the data packets can be encrypted by such improved ciphers as VEA [8]. Thus, the encrypted data volumes will be greatly reduced, and the encryption efficiency will be improved.

13.3 Secure Media Preview

Secure media preview may be used in video-on-demand services. As shown in Figure 13.3, the sender degrades the quality of the original media by perceptual encryption algorithms [9, 10], and then puts the degraded content over the web portal. Users can preview the degraded content freely. If a user is interested in

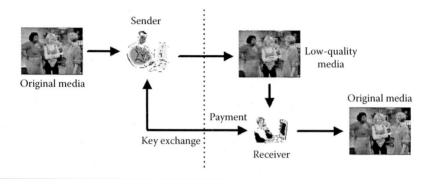

Figure 13.3   Procedures of secure media preview.

the content, he will communicate with the sender, for example, by paying for the content and receiving the key. After getting the key, he can recover the media content and see a copy with high quality. Of course, the decryption process may be online or offline, which depends on the transmission mode, that is, streaming or downloading. Because partial encryption techniques are used in perceptual encryption algorithms, the decryption efficiency is often high enough for real-time applications.

13.4   Secure Media Transcoding

With the continued development of both network and multimedia technology, the ability to watch a TV program anywhere is very nearly a reality today. It depends mainly on two techniques: the convergence of multiple networks and the scalable coding technique. Figure 13.4 shows an example based on the two techniques. First, the TV program is compressed with scalable coding. Then, the compressed data stream is transmitted through the Internet. The transcoder receives the data stream from the Internet, changes the data stream's bit rate by truncating some bits, and then sends it to mobile networks. Finally, the mobile terminal receives the data stream, decodes it, and displays the TV program with low resolution. In this scenario, the original data stream's bit rate is changed in order to adapt to the limited bandwidth of the mobile channel. Generally, the transcoder need not decompress the data stream. To realize secure transcoding, the transcoder does not know the key, and he can only truncate the encrypted data stream. Thus, a scalable encryption algorithm [11, 12] should be used here, which supports direct bit rate conversion. Additionally, partial encryption mode can be used to reduce the energy cost at the mobile terminal. For example, only the most significant layers of the scalable data stream are encrypted, while other layers are left unencrypted. Of course, the security should be confirmed by obeying the principles of secure partial encryption.

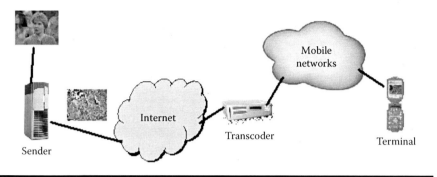

Figure 13.4   Architecture of secure media transcoding.

13.5 Secure Media Distribution

In secure media distribution, the media content is transmitted from the sender to the user in a secure manner, which protects the media content's various properties, including confidentiality and copyright. To protect the confidentiality, an encryption technique will be used. For copyright protection, watermarking or fingerprinting techniques can be adopted. In the scenario shown in Figure 13.5, the TV program is encrypted by the sender, and the authorized user can receive and decrypt the TV program. However, he cannot redistribute the decrypted program to other unauthorized users. For example, user A cannot send his copy to user B. If so, the distribution can be detected. In another case, the user cannot record the program with the capture and then send it out, for example, over the Internet. If so, the user can be traced. To realize

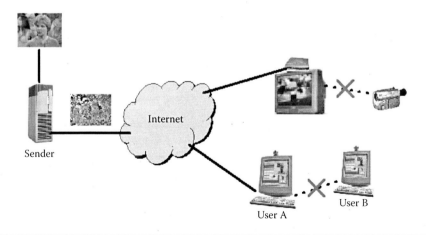

Figure 13.5   Architecture of secure media distribution.

this functionality, the commutative watermarking and encryption (CWE) scheme [13, 14] or joint fingerprint embedding and decryption (JFD) scheme [15, 16] can be used. For example, at the receiver side, during decryption, the user's unique code, for example, the set-top box ID or the user's registration code, is embedded into the TV program imperceptibly. Thus, the decrypted program contains the unique code. If the user sends it out, the unique code can be extracted and used to reveal the illegal user. If the decryption and watermarking operations are integrated in the set-top box, the computational cost is not so important. But, if they are implemented by the PC or mobile terminal, efficient algorithms are preferred.

13.6 Digital Rights Management

The role of digital rights management (DRM) [17–19] in distribution of content is to enable business models whereby the consumption and use of content is controlled. As such, DRM extends beyond the physical delivery of content into managing the content lifecycle. When a user buys the content, he may agree to certain constraints. For example, he may choose a free version for preview or a full version at cost, and for the full version, he may agree to pay a monthly fee or pay per view. DRM allows these choices to be translated into permissions and constraints, which are then enforced when the user accesses the content. Thus, to realize these

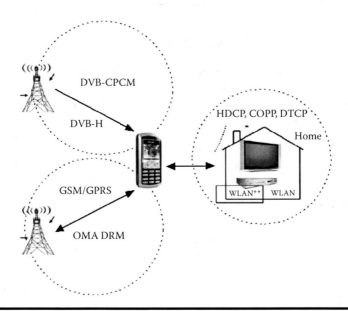

Figure 13.6   The digital rights management (DRM) systems in the proposed scenario.

functionalities, multimedia content encryption will be used. Various DRM standards have been formed, which meet different application scenarios. Figure 13.6 shows some typical DRM systems involved in the following application scenario: Bob watches a TV program with low quality on a mobile phone; when he arrives at home, he continues to watch the TV program with high quality on a home TV. This application scenario is realized based on the convergence of three networks, Digital Video Broadcast for Handset (DVB-H), Global System for Mobile communication (GSM), and Home network. For each network, there exists the corresponding DRM system.

In DVB-H network, DVB-H content protection and copy management (DVB CPCM) [17] provides the principles for content protection and copy management of commercial digital content delivered to consumer products. CPCM manages content usage from acquisition into the CPCM system until final consumption, or export from the CPCM system, in accordance with the particular usage rules of that content. Possible sources for commercial digital content include broadcast (e.g., cable, satellite, and terrestrial), Internet-based services, packaged media, and mobile services. CPCM is intended for use in protecting all types of content: audio, video, and associated applications and data.

For mobile communication, including GSM networks, Open Mobile Alliance DRM (OMA DRM) [18] defines the format and the protection mechanism for DRM Content, the format and the protection mechanism for the Rights Object, and the security model for management of encryption keys. Before the content is delivered, it is packaged to protect it from unauthorized access. A content issuer delivers DRM Content, and a rights issuer generates a Rights Object. DRM Content may be transmitted independent from Rights Objects.

In home networks, various content protection means have been reported. Digital Transmission Content Protection (DTCP) [19] is a DRM technology used to control "digital home" devices, including DVD players and televisions by encrypting their interconnections. Additionally, some Conditional Access (CA) systems [20–23] are presented to transmit TV program to TV set-top box in a secure manner.

In DRM systems, the content control or protection is realized by packaging the data into two parts, the media content and rights information, as shown in Figure 13.7. Among them, media content is first encrypted then packaged, and it can be transmitted from the sender to the receiver in a public channel. The rights information contains the content key and access rights, which is transmitted over a secret channel independent from the media content. Various media content encryption algorithms can be used to protect the media content. The selection of the encryption algorithms depends on the application's requirements, for example, security, encryption efficiency, compression efficiency, format compliance, etc.

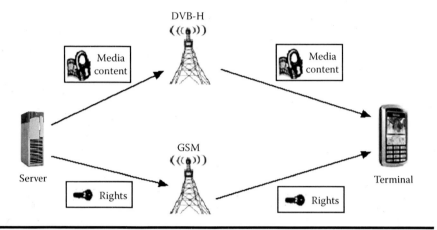

Figure 13.7 **The packaging of media content and rights information.**

References

[1] Video-on-Demand, http://en.wikipedia.org/wiki/Video-on-demand

[2] ITU IPTV Focus Group, http://www.itu.int/ITU-T/IPTV/

[3] Mobile TV, http://en.wikipedia.org/wiki/Mobile_tv

[4] ISMACryp 1.1 (ISMA Encryption & Authentication Specification 1.1). http://www.isma.tv/

[5] Secure Real-time Transport Protocol (SRTP) Security profile for Real-time Transport Protocol. IETF Request for Comments document, RFC 3711.

[6] IPSec (IP Security). http://en.wikipedia.org/wiki/IPSec.

[7] MPEG-4. http://en.wikipedia.org/wiki/MPEG4.

[8] A. S. Tosun, and W. C. Feng. 2001. Lightweight security mechanisms for wireless video transmission. In *Proceedings International Conference on Information Technology: Coding and Computing*, April 2-4, 157–161.

[9] A. Torrubia, and F. Mora. 2002. Perceptual cryptography on MPEG Layer III bitstreams. *IEEE Transactions on Consumer Electronics* 48(4): 1046–1050.

[10] S. Lian, J. Sun, and Z. Wang. 2004. Perceptual cryptography on SPIHT compressed images or videos. In *Proceedings IEEE International Conference on Multimedia and Expo (I) (ICME2004)*, Taiwan, Vol. 3, 2195–2198.

[11] S. J. Wee, and J. G. Apostolopoulos. 2003. Secure scalable streaming and secure transcoding with JPEG-2000. *Proceedings of IEEE International Conference on Image Processing*, Vol. 1, I-205–208.

[12] C. Yuan, B. Zhu, Y. Wang, S. Li, and Y. Zhong. 2003. Efficient and fully scalable encryption for MPEG-4 FGS. Paper presented at IEEE International Symposium on Circuits and Systems, Bangkok, Thailand, May 25–28.

[13] S. Lian, Z. Liu, Z. Ren, and H. Wang. 2006. Commutative watermarking and encryption for media data. *International Journal of Optical Engineering* 45(8): 0805101–0805103.

[14] S. Lian, Z. Liu, Z. Ren, and H. Wang. 2007. Commutative encryption and watermarking in compressed video data. *IEEE Circuits and Systems for Video Technology* 17(6): 774–778.

[15] S. Lian, Z. Liu, Z. Ren, and H. Wang. 2006. Secure distribution scheme for compressed data streams. Paper presented at 2006 IEEE Conference on Image Processing (ICIP 2006), Atlanta, GA, October 8–11.

[16] R. Parnes, and R. Parviainen. 2001. Large scale distributed watermarking of multicast media through encryption. In *Proceedings IFIP International Conference on Communications and Multimedia Security Issues of the New Century*, 17.

[17] Digital Video Broadcasting Content Protection & Copy Management (DVB-CPCM), DVB Document A094 Rev. 1, July 2007.

[18] Open Mobile Alliance, Digital Rights Management 2.0 (OMA DRM 2.0), 03 Mar 2006.

[19] DTCP (Digital Transmission Content Protection), http://en.wikipedia.org/wiki/DTCP

[20] T. Jiang, Y. Hou, and S. Zheng. 2004. Secure communication between set-top box and smart card in DTV broadcasting. *IEEE Transactions on Consumer Electronics* 50(3): 882–886.

[21] T. Jiang et al. 2004. Key distribution based on hierarchical access control for Conditional Access System in DTV broadcast. *IEEE Transactions on Consumer Electronics* 50: 225–230.

[22] S. Park, J. Jeong, and T. Kwon. 2006. Contents distribution system based on MPEG-4 ISMACryp in IP set-top box environments. *IEEE Transactions on Consumer Electronics*, 52(2): 660–668.

[23 F. Pescador, C. Sanz, M. J. Garrido, C. Santos, and R. Antoniello. 2006. A DSP based IP set-top box for home entertainment. *IEEE Transactions on Consumer Electronics* 52(1): 254–262.

Chapter 14

Open Issues

During the past decade, multimedia content encryption has been widely studied, and some methods have been reported, which meet various applications. However, with respect to the versatility of multimedia content and services, multimedia content encryption is still not mature, and there are some open issues in this field.

14.1 Perceptual Security

Besides cryptographic security, perceptual security is also an important metric in multimedia encryption algorithms. Perceptual security depends on the intelligibility of the encrypted media content. However, to date, no suitable objective metrics exist for the intelligibility of multimedia data. Most of the existing work uses peak signal-to-noise ratio (PSNR) to convey the degradation of media data. But PSNR is more suitable for detecting the quality than the intelligibility. For example, an image with low PSNR may be still intelligible, while a confused image may have a large PSNR. Additionally, using subjective metrics (human perception) is not efficient. Thus, suitable metrics for content intelligibility are expected.

14.2 Lightweight Implementation

With the development of multimedia and mobile communication techniques, mobile media services are becoming more and more popular. The computing capability and power storage of mobile terminals are often limited, which necessitates lightweight encryption algorithms to protect mobile media. These algorithms are time-efficient but not power-consuming. Thus, first, existing encryption algorithms with high encryption efficiency, such as partial encryption or compression-combined encryption, are preferred. Second, a method should be found to implement these algorithms on the terminal side in a lightweight manner.

14.3 Contradiction between Format Compliance and Upgrade

Format compliance is often considered when designing multimedia encryption algorithms. In this case, the format information of multimedia data is kept unchanged, and only some other sensitive information is encrypted. The format information can be used to synchronize the encoder and decoder. Thus, the encrypted media data may be displayed, and the resilience to transmission errors can be improved. According to this property, multimedia encryption algorithms should be designed according to the compression methods. That is, different compression methods need different encryption algorithms. Considering that there are various media data that have different file formats, various encryption algorithms should be designed. However, according to digital rights management (DRM) standards, all kinds of media data should be protected equally. Additionally, if one encryption algorithm is suitable for all kinds of media data, then it is easy to be implemented and upgraded in practical applications. Thus, there is a contradiction between format compliance and upgrade. The tradeoff will be decided by system designers.

14.4 Ability to Support Direct Operations

As has been mentioned in the chapter on scalable encryption, media data may be operated in the encrypted domain. Besides bit rate conversion, multimedia index may also be applied in the encrypted domain. For example, in a multimedia database, the encrypted multimedia data can be indexed without exploring the secret. It can be realized by various methods. For example, you can encrypt only the sensitive parts and leave the other part for the index. Alternatively, the key word and media content are encrypted independently, and the encrypted key word is used for the index. In practical applications, the algorithms can be designed with respect to the practical requirements. Furthermore, some other operations can be done in the encrypted domain, such as editing, object tracking, etc., which bring some new challenges to the design of multimedia content encryption.

14.5 Key Management

Key management is not considered in this book, but it is important to practical applications. For example, in different transmission modes, such as broadcast, unicast, multicast, p2p, etc., the key may be transmitted from the sender to the receiver using different methods. In certain transmission modes, different media content may be encrypted with different keys in order to control access rights to media content. For the same media content, it can be partitioned into many segments, and each segment is encrypted with a different key in order to improve the system's security. Taking the third case, for example, there are some issues that need to be solved. First, the segment size is related to the cost of subkeys. The smaller the segment size, the more subkeys need to be generated. Additionally, the segment size is related to the error-resilience. The bigger the segment size, the easier the segment is destroyed by transmission errors, without considering error-correction codes. Thus, key generation, distribution and management will be carefully designed in addition to the media encryption algorithm.

14.6 Combination between Encryption and Other Means

As is known, multimedia content encryption protects only the confidentiality of the multimedia content. Generally, in practical applications, the integrity and ownership of multimedia content are also important properties that need to be protected. Thus, such techniques as content authentication, watermarking, or fingerprinting can be considered when designing the multimedia encryption algorithms. There exist some known algorithms combining these. For example, the signcryption algorithm combines digital signature with encryption, the commutative watermarking and encryption algorithm combines watermarking and encryption, and the joint fingerprint embedding and decryption algorithm combines fingerprinting and decryption. These algorithms can meet some potential applications although they are not yet mature.

Index